STATISTICS SUCCESS
in 20 Minutes
a Day

STATISTICS
SUCCESS
in 20 Minutes
a Day

Linda J. Young

LEARNINGEXPRESS®

NEW YORK

Library of Congress Cataloging-in-Publication Data:
Young, Linda J., 1952-
    Statistics success in 20 minutes a day / Linda J. Young.
        p. cm.
    Includes bibliographical references.
    ISBN 1-57685-535-X
    1. Mathematical statistics—Problems, exercises, etc. I. Title.
QA276.2.Y68 2005
519.5—dc22   2005027521

Printed in the United States of America

9 8 7 6 5 4 3 2 1

ISBN 1-57685-535-X

For information on LearningExpress, other LearningExpress products, or bulk sales, please write to us at:
    LearningExpress
    55 Broadway
    8th Floor
    New York, NY 10006

Or visit us at:
    www.learnatest.com

About the Author ▶

Linda J. Young is professor of statistics and director of biostatistics at the University of Florida. She has previously served on the faculties at Oklahoma State University and the University of Nebraska.

Contents

Introduction ▶

If you have never taken a statistics course, and now find that you need to know the basics of statistics—this is the book for you. If you have already taken a statistics course, but felt like you never understood what the teacher was trying to tell you—this book can teach you what you need to know. If it has been a while since you have taken a statistics course, and you need to refresh your skills—this book will review the basics and reteach you the skills you may have forgotten. Whatever your reason for needing to know statistics, *Statistics Success in 20 Minutes a Day* will teach you what you need to know. It gives you the statistics basics in clear and straightforward lessons that you can do at your own pace.

▶ How to Use This Book

Statistics Success teaches basic concepts in 20 self-paced lessons. The book includes a pretest, a posttest, and tips on how to prepare for a standardized test. Before you begin Lesson 1, take the pretest to assess your current statistics abilities. The answer key follows the pretest. This will be helpful in determining your strengths and weaknesses. After taking the pretest, move on to Lesson 1.

Each lesson offers detailed explanations of a new concept. There are numerous examples with step-by-step solutions. As you proceed through a lesson, you will find tips and shortcuts that will help you learn a concept. Each new concept is followed by a set of practice problems. The answers to the practice problems are in the answer key located at the end of the book.

When you have completed all 20 lessons, take the posttest at the end of the book. The posttest has the same format as the pretest, but the questions are different. Compare the results of the posttest with the results of the pretest. What are your strengths? Do you have weak areas? Do you need to spend more time on some concepts, or are you ready to go to the next level?

▶ Make a Commitment

Success does not come without effort. If you truly want to be successful, make a commitment to spend the time you need to improve your statistics skills. When you achieve statistics success, you have laid the foundation for future challenges and opportunities.

So sharpen your pencil and get ready to begin the pretest!

STATISTICS
SUCCESS
in 20 Minutes
a Day

# Pretest

Before you begin Lesson 1, you may want to get an idea of what you know and what you need to learn. The pretest will answer some of these questions for you. The pretest consists of 50 multiple-choice questions covering the topics in this book. Although 50 questions can't cover every concept, skill, or shortcut taught in this book, your performance on the pretest will give you a good indication of your strengths and weaknesses. Keep in mind, the pretest does not test all the skills taught in this statistics book.

If you score high on the pretest, you have a good foundation and should be able to work your way through the book quickly. If you score low on the pretest, don't despair. This book will take you through the statistics concepts step by step. If you get a low score, you may need to take more than 20 minutes a day to work through a lesson. However, this is a self-paced program, so you can spend as much time on a lesson as you need. You decide when you fully comprehend the lesson and are ready to go on to the next one.

Take as much time as you need to do the pretest. You will find that the level of difficulty increases as you work your way through the pretest.

ANSWER SHEET

1. ⓐ ⓑ ⓒ ⓓ	18. ⓐ ⓑ ⓒ ⓓ	35. ⓐ ⓑ ⓒ ⓓ
2. ⓐ ⓑ ⓒ ⓓ	19. ⓐ ⓑ ⓒ ⓓ	36. ⓐ ⓑ ⓒ ⓓ
3. ⓐ ⓑ ⓒ ⓓ	20. ⓐ ⓑ ⓒ ⓓ	37. ⓐ ⓑ ⓒ ⓓ
4. ⓐ ⓑ ⓒ ⓓ	21. ⓐ ⓑ ⓒ ⓓ	38. ⓐ ⓑ ⓒ ⓓ
5. ⓐ ⓑ ⓒ ⓓ	22. ⓐ ⓑ ⓒ ⓓ	39. ⓐ ⓑ ⓒ ⓓ
6. ⓐ ⓑ ⓒ ⓓ	23. ⓐ ⓑ ⓒ ⓓ	40. ⓐ ⓑ ⓒ ⓓ
7. ⓐ ⓑ ⓒ ⓓ	24. ⓐ ⓑ ⓒ ⓓ	41. ⓐ ⓑ ⓒ ⓓ
8. ⓐ ⓑ ⓒ ⓓ	25. ⓐ ⓑ ⓒ ⓓ	42. ⓐ ⓑ ⓒ ⓓ
9. ⓐ ⓑ ⓒ ⓓ	26. ⓐ ⓑ ⓒ ⓓ	43. ⓐ ⓑ ⓒ ⓓ
10. ⓐ ⓑ ⓒ ⓓ	27. ⓐ ⓑ ⓒ ⓓ	44. ⓐ ⓑ ⓒ ⓓ
11. ⓐ ⓑ ⓒ ⓓ	28. ⓐ ⓑ ⓒ ⓓ	45. ⓐ ⓑ ⓒ ⓓ
12. ⓐ ⓑ ⓒ ⓓ	29. ⓐ ⓑ ⓒ ⓓ	46. ⓐ ⓑ ⓒ ⓓ
13. ⓐ ⓑ ⓒ ⓓ	30. ⓐ ⓑ ⓒ ⓓ	47. ⓐ ⓑ ⓒ ⓓ
14. ⓐ ⓑ ⓒ ⓓ	31. ⓐ ⓑ ⓒ ⓓ	48. ⓐ ⓑ ⓒ ⓓ
15. ⓐ ⓑ ⓒ ⓓ	32. ⓐ ⓑ ⓒ ⓓ	49. ⓐ ⓑ ⓒ ⓓ
16. ⓐ ⓑ ⓒ ⓓ	33. ⓐ ⓑ ⓒ ⓓ	50. ⓐ ⓑ ⓒ ⓓ
17. ⓐ ⓑ ⓒ ⓓ	34. ⓐ ⓑ ⓒ ⓓ	

▶ Pretest

1. The time it takes an employee to drive to work is the variable of interest. What type of variable is being observed?
   a. categorical variable
   b. continuous variable
   c. discrete variable
   d. explanatory variable

2. A study was conducted to compare two different approaches to preparing for an exam. Twenty high school students taking chemistry volunteered to participate. Ten were randomly assigned to use the first approach; the other ten used the second approach. Each one's performance on the next chemistry exam was recorded. What type of study is this?
   a. experiment with a broad scope of inference
   b. experiment with a narrow scope of inference
   c. sample survey
   d. observational study

3. Random digit dialing was used to select households in a particular state. An adult in each household contacted was asked whether the household had adequate health insurance. A critic of the poll said that the results were biased because households without telephones were not included in the survey. As a consequence, the estimated percentage of households that had adequate health insurance was biased upward. What type of bias was the critic concerned about?
   a. measurement bias
   b. nonresponse bias
   c. response bias
   d. selection bias

4. A study was conducted to determine whether a newly developed rose smelled better than the rose of the standard variety. Twenty students were randomly selected from a large high school to participate in a "smell study." Each selected student smelled both roses in a random order and selected the one that smelled best. What is the population of interest and what are the response and explanatory variables?
   a. The population is all students at the large high school; the response variable is the rose choice; and the explanatory variable is the type of rose.
   b. The population is all students at the large high school; the response variable is the type of rose; and the explanatory variable is the rose choice.
   c. The population is all roses of these two types; the response variable is the rose choice; and the explanatory variable is the type of rose.
   d. The population is all roses of these two types; the response variable is the type of rose; and the explanatory variable is the rose choice.

For problems 5 and 6, consider the following 12 data points: 10, 12, 10, 18, 16, 15, 9, 14, 11, 13, 12, and 16.

5. What is the median of these data?
   a. 9
   b. 12
   c. 12.5
   d. 13

6. What is the interquartile range of these data?
   a. 4
   b. 5
   c. 6
   d. 9

7. What does the length of the box in a boxplot represent?

   a. the range

   b. the interquartile range

   c. the median

   d. the mean

8. How does one standardize a random variable?

   a. Add the mean.

   b. Subtract the mean.

   c. Add the mean and divide by the standard deviation.

   d. Subtract the mean and divide by the standard deviation.

For problems 9 and 10, consider this information: On any given day, the probability it will rain is 0.32; the probability the wind will blow is 0.2; and the probability that it will rain and the wind will blow is 0.1.

9. For a randomly selected day, what is the probability that it will rain or the wind will blow?

   a. 0.42

   b. 0.52

   c. 0.58

   d. 0.62

10. For a randomly selected day, what is the probability that it will NOT rain and the wind will NOT blow?

   a. 0.38

   b. 0.48

   c. 0.58

   d. 0.90

Use the following information for problems 11, 12, and 13. The students in a small high school were surveyed. Each student was asked whether he or she used a safety belt whenever driving. This information and the gender of the student was recorded as follows:

USE OF SAFETY BELTS

	USE SAFETY BELT?		
GENDER	YES	NO	TOTALS
Female	82	23	105
Male	69	31	100
Totals	151	54	205

11. What is the probability that a randomly selected student is a male who does not use his seat belt?

   a. $\dfrac{31}{100}$

   b. $\dfrac{23}{105}$

   c. $\dfrac{31}{205}$

   d. $\dfrac{54}{205}$

12. What is the probability that a randomly selected student is a female given that the person is a seat belt user?

   a. $\dfrac{82}{105}$

   b. $\dfrac{82}{151}$

   c. $\dfrac{82}{205}$

   d. $\dfrac{151}{205}$

13. Is the use of a safety belt independent of gender?
   a. no, because the probability that a randomly chosen student is a female does not equal the probability of female given safety belt use
   b. no, because the number of females who use a safety belt is not equal to the number of males who use a seat belt
   c. yes, because the sample was randomly selected
   d. yes, because both genders *do* use seat belts more than they do *not* use seat belts

For problems 14 and 15, consider that 1% of a population has a particular disease. A new test for identifying the disease has been developed. If the person has the disease, the test is positive 94% of the time. If the person does not have the disease, the test is positive 2% of the time.

14. What is the probability that a randomly selected person from this population tests positive?
   a. 0.0292
   b. 0.0094
   c. 0.096
   d. 0.96

15. A person is randomly selected from this population and tested. She tests positive. Which of the following best represents the probability that she has the disease?
   a. 0.0094
   b. 0.32
   c. 0.34
   d. 0.94

Use the following information for problems 16, 17, and 18. On any given day, the probability that Megan will be late for work is 0.2. Whether or not she is late to work is independent from day to day.

16. Megan was late to work today. What is the probability that she will NOT be late to work tomorrow?
   a. 0.16
   b. 0.2
   c. 0.6
   d. 0.8

17. Which of the following is closest to the probability that Megan will be late to work at least one of the five days next week?
   a. 0.00032
   b. 0.33
   c. 0.41
   d. 0.67

18. What is the probability that Megan will be on time exactly three days and then be late on the fourth one?
   a. 0.0064
   b. 0.1024
   c. 0.16
   d. 0.512

19. A train is scheduled to leave the station at 3 P.M. However, it is equally likely to actually leave the station any time from 2:55 to 3:15 P.M. What is the probability it will depart the station early?
   a. 0.25
   b. 0.33
   c. 0.67
   d. 0.75

20. Let z be a standard normal random variable. Find the probability that a randomly selected value of z is between -2.1 and 0.4.
   a. 0.1079
   b. 0.3446
   c. 0.5475
   d. 0.6554

21. Let z be a standard normal random variable. Find z^* such that the probability that a randomly selected value of z is greater than z^* is 0.2.
   a. -0.84
   b. 0.4207
   c. 0.5793
   d. 0.84

22. Let X be a normal random variable with mean 20 and standard deviation 5. What is the probability that a randomly selected value of X is between 15 and 25?
   a. 0.32
   b. 0.68
   c. 0.95
   d. 0.997

23. A random sample of size 25 is selected from a population that is normally distributed with a mean of 15 and a standard deviation of 4. What is the sampling distribution of the sample mean?
   a. normal with a mean of 0 and a standard deviation of 1
   b. normal with a mean of 15 and a standard deviation of 0.16
   c. normal with a mean of 15 and a standard of 0.8
   d. normal with a mean of 15 and a standard deviation of 4

24. Find t^* such that the probability that a randomly selected observation from a t-distribution with 16 degrees of freedom is less than t^* is 0.1.
   a. -1.746
   b. -1.337
   c. 1.337
   d. 1.746

25. A researcher decides to study the bite strength of alligators. She believes that if she takes a large enough random sample, she will be able to say that the average of the bite strengths she records will be close to the mean bite strength of all the alligators in the population she is studying. Is she correct?
   a. No. One can never be sure that the sample mean is close to the population mean.
   b. Yes. By the Central Limit Theorem, the sample mean will be equal to the population mean if $n > 30$.
   c. Yes. By the Central Limit Theorem, the sample mean will be approximately normally distributed, and the mean of the sampling distribution will be the population mean.
   d. Yes. By the Law of Large Numbers, as the sample size increases, the sample mean will get close to the population mean.

26. A poll was conducted to determine what percentage of the registered voters favored having the current mayor of a large city run for a second term. The results were that 52% of the registered voters polled were in favor of the second term with a margin or error of 0.04. What does this mean?
   a. There is only a 4% chance that 52% of the registered voters do not favor the mayor running for a second term.
   b. The estimated percentage of 52% is sure to be within 4% of the true percentage favoring the mayor's run for a second term.
   c. With 95% confidence, the estimated percentage of 52% is within 4% of the true percentage favoring the mayor's run for a second term.
   d. With 96% confidence, the estimated percentage of 52% is equal to the true percentage favoring the mayor's run for a second term.

27. A national organization wanted to estimate the proportion of adults in the nation who could read at the eighth grade level or higher. The organization decided that it would be nice to have estimates for each state as well. To accomplish this, they took a random sample of adults within each state and combined the data to obtain a national estimate. What type of sampling plan is this?

a. cluster sampling

b. simple random sampling

c. stratified random sampling

d. systematic random sampling

28. A researcher was studying greenbugs on oats. To obtain an estimate of the population in a large oat field, he picked a random point in the field and a random direction. He went to the random starting point and counted the number of greenbugs on the oat plant closest to that point. He then took ten steps in the direction that had been selected at random and counted the number of greenbugs on the closest oat plant. He repeated the process of taking ten steps in the same direction and counting the number of greenbugs on the closest oat plant until he had counted the number of greenbugs on 50 plants. What type of sampling plan is this?

a. cluster sampling

b. simple random sampling

c. stratified random sampling

d. systematic random sampling

29. A researcher set a 95% confidence interval on the mean length of fish in a recreational lake and found it to be from 6.2 to 8.7 inches. Which of the following is a proper interpretation of this interval?

a. Of the fish in the recreational lake, 95% are between 6.2 and 8.7 inches long.

b. We are 95% confident that the sample mean length of fish in the recreational lake is between 6.2 and 8.7 inches.

c. We are 95% confident that the population mean length of fish in the recreational lake is between 6.2 and 8.7 inches.

d. There is a 95% chance that a randomly selected fish from the recreational lake will be between 6.2 and 8.7 inches.

30. A large university wanted to know whether toilet paper in the campus restrooms should be hung so that the sheets rolled off over the top or under the bottom of the roll. Two hundred students were randomly selected to participate in a survey. Each selected student was asked his or her preference on how to hang the toilet paper. Researchers found 66% preferred that the sheets roll over the top of the roll. Which of the following would be used to set a 95% confidence interval on the proportion of this university's student population favoring the sheets to roll off the top?

a. $0.66 \pm 1.645 \dfrac{\sqrt{0.66(1-0.66)}}{200}$

b. $0.66 \pm 1.96 \sqrt{0.66(1-0.66)}$

c. $0.66 \pm 1.96 \sqrt{\dfrac{0.66(1-0.66)}{200}}$

d. $0.66 \pm 1.645 \sqrt{0.66(1-0.66)}$

31. An owner of a swimming pool wants to know whether or not she needs to add chlorine to the water. Because of costs and the fact that too much chlorine is unpleasant for swimmers, she wants to be sure that chlorine is needed before adding it. What is the swimming pool owner's null hypothesis and what would be a type I error?

a. H_0: No additional chlorine is needed in the pool. Type I error would occur if she added chlorine when it was not needed.

b. H_0: No additional chlorine is needed in the pool. Type I error would occur if she did not add chlorine when it was needed.

c. H_0: Additional chlorine is needed in the pool. Type I error would occur if she added chlorine when it was not needed.

d. H_0: Additional chlorine is needed in the pool. Type I error would occur if she did not add chlorine when it was needed.

32. A consumer advocate group doubted the claim of a candy company that 30% of the packages had a price. They believed the percentage of packages with prizes was much less. Let p be the true proportion of this candy maker's packages that have a prize, and let $\widehat{p}$ be the sample proportion of those packages with a prize. What is the appropriate set of hypotheses for the consumer advocate group to test?

a. H_0: $p = 0.3$; H_a: $p < 0.3$

b. H_0: $p = 0.3$; H_a: $p \neq 0.3$

c. H_0: $\widehat{p} = 0.3$; H_a: $\widehat{p} < 0.3$

d. H_0: $\widehat{p} = 0.3$; H_a: $\widehat{p} \neq 0.3$

33. Nationally, the percentage of people aged 12 to 54 with myopia (near sightedness) has been reported to be 25%. A researcher believes that a higher percentage of people in the same age group is myopic in her area. To test this assumption, she selects a random sample of people in the region and determines whether each person is myopic. Of 200 people surveyed, 56 are myopic. What is the appropriate test statistic to test the researcher's hypothesis?

a. $z_T = \dfrac{0.28 - 0.25}{\sqrt{0.25\dfrac{(1 - 0.25)}{200}}}$

b. $z_T = \dfrac{0.28 - 0.25}{\sqrt{0.28\dfrac{(1 - 0.28)}{200}}}$

c. $z_T = \dfrac{0.28 - 0.25}{\dfrac{\sqrt{0.25(1 - 0.25)}}{200}}$

d. $z_T = \dfrac{0.28 - 0.25}{\dfrac{\sqrt{0.28(1 - 0.28)}}{200}}$

34. A statistician was testing the following set of hypotheses: H_0: $p = 0.86$ versus H_a: $p \neq 0.86$. Using a random sample of size 178, he found $z_T = 1.58$. What is the p-value associated with this test?

a. 0.0285

b. 0.0571

c. 0.1142

d. 0.9429

35. A statistician conducted a hypothesis test and found the p-value to be 0.04. Using a 5% level of significance, what conclusion should she make?

a. Accept the null hypothesis.

b. Do not reject the null hypothesis.

c. Reject the alternative hypothesis.

d. Reject the null hypothesis.

Use the following information for problems 36 and 37. A forester wanted to determine the mean height of trees in an area that was planted with trees a number of years ago. He selected 36 trees at random and determined the height of each. The sample mean was 54.2 meters, and the sample standard deviation was 8.1 meters.

36. What is the appropriate multiplier to use in setting a 95% confidence interval on the mean height of trees in the study region?

   a. 1.690

   b. 1.96

   c. 2.028

   d. 2.030

37. Given the proper multiplier, which of the following represents a 95% confidence interval on the mean height of trees in this area?

   a. $54.2 \pm multiplier \times 8.1$

   b. $54.2 \pm multiplier \times \sqrt{\dfrac{8.1}{36}}$

   c. $54.2 \pm multiplier \times \dfrac{8.1}{\sqrt{36}}$

   d. $54.2 \pm multiplier \times \dfrac{8.1}{36}$

38. A lightbulb manufacturer wanted to be sure that her bulbs burned, on average, longer than the 1,500 hours advertised. She selected a random sample of 56 bulbs and measured the time it took for each to burn out. The sample mean was 1,512 hours, and the sample standard deviation was 38 hours. What is the appropriate test statistic to test the hypothesis that the manufacturer is interested in?

   a. $t_T = \dfrac{1{,}512 - 1{,}500}{\dfrac{38}{\sqrt{56}}}$

   b. $t_T = \dfrac{1{,}512 - 1{,}500}{38}$

   c. $t_T = \dfrac{1{,}500 - 1{,}512}{\dfrac{38}{\sqrt{56}}}$

   d. $t_T = \dfrac{1{,}500 - 1{,}512}{38}$

39. A statistician conducts a test of the following set of hypotheses: $H_0: \mu = 18$ versus $H_a: \mu \neq 18$. Based on a random sample of 42, she found the value of the test statistic to be 1.78. What is the p-value associated with the test?

   a. $0.0125 < p < 0.025$

   b. $0.025 < p < 0.05$

   c. $0.05 < p < 0.10$

   d. $p = 0.0375$

40. A botanist believes that he has developed a new variety of rose that, on average, has more blooms on each plant than the standard variety for the area. He randomly selects 32 plants from his new variety and 32 plants from the standard variety. The plants are each properly cared for throughout the growing season and the total number of blooms each produces is recorded. Statistically, what does the researcher want to do?

   a. Compare two treatment means using a matched-pairs design.
   b. Compare two treatment means using a two-group design.
   c. Compare the means from two populations.
   d. Compare the means of two samples from the same population.

41. In a two-group design, 32 observations were taken under the first treatment, and 36 were taken under the second treatment. The sample variance under the first treatment was 6.7 and under the second was 5.9. Believing that both of these are estimates of a common variance, the statistician wants to obtain an estimate of this common variance. How should that be done?

   a. $s_p^2 = \dfrac{6.7 + 5.9}{2}$

   b. $s_p^2 = \dfrac{32(6.7) + 36(5.9)}{68}$

   c. $s_p^2 = \dfrac{31(6.7) + 35(5.9)}{66}$

   d. cannot be determined because the sample sizes are not the same for the two treatments

Use the following information for problems 42 and 43. A study has been conducted using a two-group design. Fifty-four units received the first treatment, and 47 received the second treatment. Based on theory, the researcher believes that the variances under the two treatments will be different. The sample standard deviation under the first treatment is 3.4 and under the second is 16.7.

42. What is the standard error of the estimated difference in treatment means $(\overline{X}_1 - \overline{X}_2)$?

   a. $\sqrt{\dfrac{3.4}{54} + \dfrac{16.7}{47}}$

   b. $\dfrac{3.4}{\sqrt{54}} + \dfrac{16.7}{\sqrt{47}}$

   c. $\dfrac{3.4}{54} + \dfrac{16.7}{47}$

   d. $\sqrt{\dfrac{3.4^2}{54} + \dfrac{16.7^2}{47}}$

43. Approximately how many degrees of freedom are associated with the standardized estimate of the difference in treatment means $(\overline{X}_1 - \overline{X}_2)$?

   a. $54 + 47 - 2 = 99$

   b. $54 + 47 = 101$

   c. $\dfrac{\left(\dfrac{3.4}{54} + \dfrac{16.7}{47}\right)^2}{\dfrac{1}{53}\left(\dfrac{3.4}{54}\right)^2 + \dfrac{1}{46}\left(\dfrac{16.7}{47}\right)^2}$

   d. $\dfrac{\left(\dfrac{3.4^2}{54} + \dfrac{16.7^2}{47}\right)^2}{\dfrac{1}{53}\left(\dfrac{3.4^2}{54}\right)^2 + \dfrac{1}{46}\left(\dfrac{16.7^2}{47}\right)^2}$

Use the following information for problems 44, 45, and 46. A study has been conducted using a matched-pairs design. Forty pairs were used in the study. The sample standard deviation under the first treatment is 6.5, and the sample standard deviation under the second treatment is 8.6. The standard deviation of the differences within each pair is 12.2.

44. What is the standard error of the estimated difference in treatment means $(\overline{X}_1 - \overline{X}_2)$?

a. $\sqrt{\dfrac{6.5}{40} + \dfrac{8.6}{40}}$

b. $\sqrt{\dfrac{6.5^2}{40} + \dfrac{8.6^2}{40}}$

c. $\sqrt{\dfrac{12.2}{40}}$

d. $\sqrt{\dfrac{12.2^2}{40}}$

45. How many degrees of freedom are associated with the standardized estimate of the difference in treatment mean $(\overline{X}_1 - \overline{X}_2)$?

a. $40 - 1 = 39$

b. $40 + 40 - 2 = 78$

c. 40

d. $\dfrac{\left(\dfrac{6.5^2}{40} + \dfrac{8.6^2}{40}\right)^2}{\dfrac{1}{39}\left(\dfrac{6.5^2}{40}\right)^2 + \dfrac{1}{39}\left(\dfrac{8.6^2}{40}\right)^2}$

46. A chi-squared goodness-of-fit test was conducted. There were seven categories, but no parameters were estimated. The value of the test statistic was 9.6. What is the p-value associated with the test?

a. $0.05 < p < 0.075$

b. $0.10 < p < 0.15$

c. $0.15 < p < 0.20$

d. $0.20 < p < 0.30$

Use the following information for problems 47 and 48. A high school counselor wanted to know whether there is a relationship between a student's work outside of school and his or her participation in band. He selected a random sample of 75 students. Each selected student was asked whether he or she worked full or part time during the school year and whether he or she participated in band. The results are in the following table.

BAND MEMBERS WHO WORK

WORK?	BAND?		
	YES	NO	TOTALS
Yes	5	24	29
No	14	32	46
Totals	19	56	75

47. What type of test is to be conducted?

a. paired t-test

b. chi-squared goodness-of-fit test

c. chi-squared test of homogeneity

d. chi-squared test of independence

48. How many degrees of freedom are associated with the test?

   a. 1
   b. 2
   c. 3
   d. 4

Use the following information for the final two problems. A homemaker wanted to know which of two furniture polishes worked best. One week, he decided to use one of the furniture polishes on half of each piece of furniture and to use the other furniture polish on the other half. To decide which half got the first furniture polish, he flipped a coin. A head indicated that the first furniture polish went on the left side; a tail indicated that the first furniture polish went on the right side. Thirty-three pieces of furniture were treated in this fashion. At the end of the week, he asked a friend to rate the appearance of each side on a scale of 1 to 10. This was repeated for all pieces of furniture.

49. What is the appropriate statistical test for the hypotheses of interest?

   a. z-test
   b. paired t-test
   c. independent t-test
   d. chi-squared test of independence

50. The first furniture polish scored higher, on average, than the second furniture polish. The value of the test statistic was 1.9. What is the p-value of the test, and what decision would one make concerning the hypotheses using a 5% significance level?

   a. $0.25 < p < 0.05$. Do not reject the null hypothesis.
   b. $0.25 < p < 0.05$. Reject the null hypothesis.
   c. $0.05 < p < 0.1$. Do not reject the null hypothesis.
   d. $0.05 < p < 0.1$. Reject the null hypothesis.

► **Answers**

1. b		**26.** c	
2. b		**27.** c	
3. d		**28.** d	
4. c		**29.** c	
5. c		**30.** c	
6. b		**31.** a	
7. b		**32.** a	
8. d		**33.** a	
9. a		**34.** c	
10. c		**35.** d	
11. c		**36.** d	
12. b		**37.** c	
13. a		**38.** a	
14. a		**39.** c	
15. b		**40.** c	
16. d		**41.** c	
17. d		**42.** d	
18. b		**43.** d	
19. a		**44.** d	
20. c		**45.** a	
21. d		**46.** b	
22. b		**47.** d	
23. c		**48.** a	
24. b		**49.** b	
25. d		**50.** c	

1 ▶ Populations, Samples, and Variables

LESSON SUMMARY

Numerical information permeates our lives. The morning weather report forecasts the chance of rain, and we make a decision as to whether or not to take an umbrella. Given the latest study results on the health risks of over-the-counter painkillers, we decide whether to take something to reduce the pain from a sore knee. A friend wants to attend a very selective university and wonders whether an SAT score of 1,400 or higher will ensure her admittance. A neighbor was told that there was a peculiar shadow on an X-ray and must decide whether to have a biopsy taken. The stock market has had several days of losses, and an investor wonders whether this trend will continue. Our understanding of these and many other issues will be deeper as we learn more about the discipline of statistics. But what is *statistics*? Learning the answer to this question, as well as some fundamental terms in statistics, such as *population*, *sample*, and *variable*, is the focus of this lesson.

▶ Populations and Samples

Statistics is the science of collecting, analyzing, and drawing conclusions from data. This process of collecting, analyzing, and drawing conclusions begins with the desire to answer a question about a specific population. In statistics, a *population* is the collection of individuals or objects of interest. These individuals or objects may be referred to as *members* or *units* of the population. If we are able to record all desired information on each unit in the population, then we have *taken a census*. The problem is that we rarely have the ability to gather the information from every unit of the population, due to financial constraints, time limitations, or some other reason. We must be satisfied with observing the information for only a *sample*, or a *subset* of the population of interest.

Care must be taken in obtaining the sample if we are to be able to draw solid conclusions from it. For example, if we are interested in whether a majority of the voters in a particular state would favor increasing the minimum driving age, then we would not want to simply call several households and ask the person answering the phone whether he or she favored increasing the minimum driving age. In households with children, the children are more likely to answer the phones than the adults, and the views of these non-voters might be quite different from their voting parents. Deciding how to select the sample from the population is an important aspect of data collection.

Example

A proposal before a state's legislature would increase the gasoline tax. The additional funds would be used to improve the state's roads. Some state legislators are concerned about how the voters view this proposal. To gain this information, a pollster randomly selects 1,009 registered voters in the state and asks each whether or not he or she favors the additional tax for the designated purpose. Describe the population and sample.

Solution

The population is all registered voters in the state. The sample is made up of the 1,009 registered voters who were polled.

▶ Practice

1. The leaders of a large company want to know whether on-site day care would be considered a valuable employee benefit. They randomly selected 200 employees and asked their opinion about on-site day care. Describe the population and sample.

2. A farmer saw some bollworms in his cotton field. Before deciding whether or not to reduce the number of bollworms by spraying his field, he selected 50 plants and checked each carefully for bollworms. Describe the population and sample.

3. A manufacturer received a large shipment of bolts. The bolts must meet certain specifications to be useful. Before accepting shipment, 100 bolts were selected, and it was determined whether or not each met specifications. Describe the population and sample.

▶ Types of Variables

There are two primary branches of statistics: descriptive statistics and inferential statistics. Once data have been collected or an appropriate data source identified, the information should be organized and summarized. Tables, graphs, and numerical summaries allow increased understanding and are efficient ways to present the data. *Descriptive statistics* is the branch of statistics that focuses on summarizing and displaying data.

Sometimes, description alone is not enough. People want to use data to answer questions or to evaluate decisions that have been made. *Inferential statistics* is the branch of statistics that uses the information gathered from a sample to make statements about the population from which it was selected. Because we have seen only a portion of the population (the sample), there is a chance that an incorrect conclusion can be made about the population. One role of statistics is to quantify the chance of making an incorrect conclusion.

If every population unit (person or object) were identical, no need would exist for statistics. For example, if all adult men in the United States were exactly the same number of inches tall, we could measure the height of one adult male in the United States and then know exactly how tall all U.S. adult males are. Obviously, that will not work. The heights of men vary. Some are taller than 70 inches; some are shorter than 70 inches; a small proportion is 70 inches tall. It is this variability in heights that makes determining height characteristics about the population of U.S. adult males a statistical challenge.

Every person or object in a given population typically has several characteristics that might be studied. Suppose we are interested in studying the fish in a lake. The length, weight, age, gender, and the level of methyl mercury are but a few of the characteristics that could be recorded from each. A *variable* is a characteristic that may be recorded for each unit in the population, and the observed value of the variable is generally not the same for all units. The length, weight, age, gender, and mercury level of the fish are five variables, some or all of which might be of interest in a particular study.

Data consist of making observations on one or more variables for each sampled unit. A *univariate data set* consists of observations collected regarding only one variable from each unit in the sample or population. A *bivariate data set* results from observations collected regarding two variables from each unit in the sample or population. When observations are collected on three or more variables, then we have a *multivariate data set*. (Sometimes, bivariate data sets are called multivariate data sets. Because *multi* implies more than one, this is an acceptable use of the term.)

When working with bivariate or multivariate data, the variables may have different uses. For the fish data, the goal of the study may be to predict the level of methyl mercury in fish; that is, methyl mercury level is the response variable. A *response variable,* or *outcome variable,* is one whose outcome is of primary interest. The methyl mercury level could depend on many factors, including the environment and traits of the fish. Fish length, age, weight, and gender may be potentially useful in explaining the level of methyl mercury and are called explanatory variables. An *explanatory variable* is one that may explain or cause differences in the response variable.

Notice, in the fish example, that the natures of the variables differ. Length, weight, and age are *numerical* (or *quantitative*) *variables*; that is, each observation for these variables is a number. A numerical variable is said to be continuous if the set of possible values that may be observed for the variable has an uncountable number of points; that is, the set of possible values of the variable includes one or more intervals on the number line. Length and weight represent two continuous variables. Both must be positive. Although the sensitivity of the measuring device may limit us to recording observations to the nearest millimeter or gram, the true values could be any value in an interval.

A *numerical value* is said to be discrete if the set of possible values that may be observed for this variable has a countable number of points. The ages of fish are often determined by growth rings on the scales. In the summer, fish grow rapidly, forming a band of widely separated, light rings. During the winter, slower growth is indicated by narrow separations between the rings, resulting in a dark band. Each pair of rings indicates one year. Because fish spawn at a specific time of year, during the spring for many species, age is generally recorded by year. Age 0 fish are less than a year old, age 1 fish are between 1 and 2 years, etc. Thus, age is a discrete variable with possible values of 0, 1, 2, . . .

Although there is undoubtedly an upper limit to age, we have represented the possible ages as being a countably infinite number of values.

Gender is a different type of variable; it is categorical (or qualitative) in nature. A *variable* is categorical if the possible responses are categories. Each fish is in one of two categories: male or female. We may arbitrarily associate a number with the category, but that does not change the nature of the variable. Car manufacturers, brands of battery, and types of injury are other examples of categorical variables.

▶ Practice

For each of the following, specify whether the variable is numerical or categorical. For those that are numerical, indicate whether they are discrete or continuous.

4. a person's dominant hand
5. the distance a car travels on a tank of gas
6. the color of a person's eyes
7. the number of credit card transactions at a grocery store during a 24-hour period
8. the amount of coffee dispensed by a particular machine
9. the manufacturer of the next car through a specified intersection
10. the number of students in a class of 56 who turn in a term project early

▶ In Short

This lesson has provided a brief overview of some of the key ideas in statistics. As with any science, terms have special meaning, and a number of the common statistical terms have been introduced in this lesson. Both the ideas and terms will be encountered frequently throughout this text, helping you to become more comfortable with them.

Studies

LESSON SUMMARY

Statistics involves the collection and analysis of data. Both tasks are critical. If data are not collected in a sensible manner, no amount of sophisticated analysis will compensate. Similarly, improper analyses can result in improper conclusions from even the best data. A key to a successful study is to establish a solid framework. In this lesson, we will outline such a framework and discuss the types of inference that can be made from different types of studies.

▶ Steps in Planning and Conducting Studies

Most studies are undertaken to answer one or more questions about our world. Would drilling for oil and gas in the Arctic National Wildlife Refuge negatively affect the environment? Do laws mandating seat belt use increase the rates of their use? Is the flu vaccine safe and effective in preventing illness? These are the types of questions statisticians like to answer.

Planning and conducting a study can be outlined in five steps, each of which we will discuss briefly:

1. developing the research question
2. deciding what to measure and how to measure it
3. collecting the data
4. analyzing the data
5. answering the question

Developing the Research Question

Statisticians often work in teams with other researchers. The team works together to determine the research question to be addressed in an upcoming study. To fully specify the research question, the study population should be identified and the goals of the study should be outlined. The statistician must understand the question(s) and the goals of the study if he or she is to be a full member of this team.

Deciding What to Measure and How to Measure It

Once the research question has been specified, the team must determine what information is needed to answer the research question. Identifying what variables will be measured and deciding how they will be measured is fundamental. Sometimes, this step is obvious (as in a study relating salaries of individuals to educational level). At other times, this is extremely challenging (as in a study relating attitudes toward school to intelligence).

In some studies, a comparison of two or more regimens or procedures may be the focus of the research question. As an illustration, a study could be used to determine whether students perform better on English tests if they study in a quiet environment or while listening to classical music. To answer the question, some students would study in a quiet environ-

ment; others would study while listening to classical music. The scores on the English test for each group would be used to answer the research question. The study environments (quiet or classical music) would be the treatments in this study. A *treatment* is a specific regimen or procedure assigned to the participants of the study.

Collecting the Data

Good data collection is a crucial component of any study. Because resources are always limited, the first question is whether an existing data source exists that could be used to answer the research question. If existing data are found, the manner in which the data were collected and the purpose for which they were collected must be carefully considered, so that any resulting limitations they would impose on the proposed study can be evaluated and judged to be acceptable. If no existing data are found, a careful plan for data collection must be prepared. The manner in which data are collected determines the appropriate statistical analyses to be conducted and the conclusions that can be drawn.

Analyzing the Data

Before data are collected, the analysis should be outlined. With the analysis and potential conclusions in mind, the research question should be reviewed to confirm that the planned study has the potential of answering the question. Too often, studies are conducted before the researchers realize they have no idea how to analyze the data or that the collected data cannot be used to answer the research question. The statistician should verify that the data collection protocol was properly followed. Each analysis should begin by summarizing the data graphically and numerically. Then the appropriate statistical analyses should be conducted.

Answering the Question

Through interpretation of the analysis results, we learn what conclusions can be drawn from the study. The aim is to answer the research question using the conclusions drawn from the study. Sometimes, we are unable to answer the question or are able to only partially answer it. At the conclusion of any study, the research team should reflect on what was learned from the study and use that to direct future research.

▶ Selecting the Sample

Most of the inferential methods introduced in the text are based on random selection. The simplest form of random selection is simple random sampling. A *simple random sample* of size *n* is one drawn in such a manner that every possible sample of size *n* has an equal chance of being chosen.

It is important to realize that, if every unit in the population has an equal chance of being included in a sample, the sample may still not be a simple random one. To see this, suppose that a company has two divisions, A and B. There are 700 employees in division A and 300 in division B. The management decides to take a sample of 100 employees. To do this, they write each employee's name on a chip and put the chip in bowl A or B, depending on whether the employee is in division A or division B, respectively. The chips are thoroughly mixed in each bowl. Seventy chips are drawn from bowl A and 30 chips are drawn from bowl B, and the employees whose names are on the selected chips comprise the sample. Each employee has a 1 in 100 chance of being included in the sample; however, this is not a random sample.

Only samples with 70 division-A employees and 30 division-B employees are possible; it would not be possible to have, for instance, a sample with 50 division-A employees and 50 division-B employees. Because not all samples of size 100 are equally likely to be selected,

this is not a random sample. In Lesson 14, we will discuss other methods of random selection.

Care must be taken in selecting a sample so that it is not biased. *Bias* is the tendency for a sample to differ in some systematic manner from the population of interest. Some common sources of bias are selection bias, measurement bias, response bias, and nonresponse bias. *Selection bias* occurs when a portion of the population is systematically excluded from the sample. For example, suppose a company wants to estimate the percentage of adults in a community who smoke. If a telephone poll is conducted, adults without telephones would be excluded from the sample, and selection bias would be introduced.

Measurement bias, or *response bias,* occurs when the method of observation tends to produce values that are consistently above or below the true value. For example, if a scale is inaccurately calibrated, observed weights could be consistently greater than true weights, resulting in a measurement bias. The way in which a survey question is worded could influence the response, leading to bias. For example, suppose that a survey question was stated as follows: "Many people think driving motorcycles is dangerous. Do you agree?" When stated in this way, the proportion of those agreeing will tend to be larger than would have been the case if the question had been phrased in a neutral way. The tendency of people to lie when asked about illegal behavior or unpopular beliefs, characteristics of the interviewer, and the organization taking the poll could be other sources of response bias.

Often in surveys, some people refuse to respond. *Nonresponse bias* is present if those who respond differ in important ways from those who do not participate in the survey. In a survey of gardeners, those with smaller gardens were much more likely to respond than those with large gardens. Because some of the questions were related to the size of the garden, this nonresponse resulted in response bias.

▶ Practice

1. The Chamber of Commerce of a certain city was planning a campaign to encourage the city's residents to shop within the city. Before beginning the campaign, the Chamber decided to determine how often people were shopping outside of the city. They conducted a telephone survey. Telephone numbers of households in the city were randomly selected (how this might be done will be discussed in Lesson 14). The selected numbers were called until 400 responses were obtained. An adult at each household in which the phone was answered was asked, "How many times in the past week have you shopped outside of this city?" Give an example of each of the following potential sources of bias and describe how it might affect the estimate of the average number of times that a person in this city has shopped elsewhere during the past week.

a. selection bias

b. response bias

c. nonresponse bias

2. An ornithologist (one who studies birds) wants to determine the kinds and numbers of birds that inhabit an area along an abandoned railroad track that is to become a walking path. Each morning, she goes to a randomly selected point along the track and walks 50 yards north, identifying the kinds of birds and the numbers of each kind within 10 yards of the track. She is able to clearly see all birds on, or right by, the track, but because of the tall grass and shrubs, it becomes more difficult to see birds the farther they are from the track. She then determines the average numbers of each kind within a ten square-yard area. What type of bias might she encounter and how might it affect the estimates?

▶ Types of Studies

In determining what types of conclusions can be made from a study, two primary considerations are (1) how the units are selected for inclusion in the study and (2) how the treatments are assigned to the units.

If the study units are selected at random from a population, then inference can be made to the population from which the units were drawn. Inference can only be drawn to the units in the study if units were not randomly selected from some population. If a researcher gets volunteers to participate in a study, then conclusions can be made only for those volunteers. Often, an effort is made to argue that the units in the study are representative of some larger population. However, if someone disagrees with the results and claims that the units in the study are different and that this affected the outcome, then there is no statistical foundation upon which we could argue otherwise.

In some studies, treatments can be assigned at random to units. For example, the treatments could be a new type and a standard type of dog food. Half of the dogs available for the study could be randomly assigned to the new type of dog food; the other half would get the standard type. If the assignment is made at random, then the study is called an experiment, and thus, a cause-and-effect relationship can be claimed. Returning to the dog food study, if the dogs on the new type of dog food had improved health compared to those on the standard dog food, we could conclude that the type of food caused the difference. (More advanced methodology may be used to derive casual relationships, but they are beyond the scope of this book.)

If treatments are not assigned at random to the study units, then we can discuss associations but not cause and effect. For a long time, it was observed that people who smoked were more likely to develop lung cancer than those who did not smoke. However, smoking is not a treatment that can be randomly assigned (at least ethically) to people. Therefore, it could not be claimed that smoking caused cancer, only that the two were associated with each other.

If treatments are assigned at random to units that were randomly selected from a population of interest, then the study is an experiment with a *broad scope of inference*. The term *broad scope of inference* means that inference can be drawn beyond the study units to the whole population.

If treatments are assigned at random but the units were not randomly selected from some population, then the study is an experiment with a *narrow scope of inference*. Because no random selection of the study units occurred, inference can be made only to the units in the study, and this is called a *narrow scope of inference*.

If treatments are not assigned at random but units are randomly selected from a population of interest, then the study is a *sample survey*. Notice that when conducting a survey, it is not possible to assign certain treatments at random. Gender, age, and dominant hand are only three examples. Associations, but not cause-and-effect conclusions, can be concluded for the population.

If treatments are not assigned at random and units are not randomly selected from a population of interest, then the study is an *observational study*. Here, associations can be drawn, but only for units in the study.

The discussion in the previous paragraphs is summarized in Table 2.1. To illustrate using Table 2.1, consider the following study. Suppose the goal is to determine whether the nicotine patch increases the proportion of heavy smokers (those smoking at least a pack a day) who are able to stop smoking. The study could be conducted in several ways. First, suppose that an advertisement is placed in a newspaper asking heavy smokers who want to quit smoking to participate in a study. All interested participants are screened to be sure that they are heavy smokers and to confirm a genuine interest in quitting. Half of

Table 2.1 Treatment and selection

	TREATMENTS ASSIGNED RANDOMLY TO UNITS	TREATMENTS NOT ASSIGNED RANDOMLY TO UNITS	INFERENCES THAT CAN BE DRAWN
Units Are Randomly Selected from Population for Inclusion in the Study	Experiment with broad scope of inference	Sample survey	Inference can be drawn to the population from which the units were randomly selected
Units Are NOT Randomly Selected from Population for Inclusion in the Study	Experiment with a narrow scope of inference	Observational study	Inference can be drawn only to the unit used in the study
Inferences That Can Be Drawn	Cause-and-effect conclusions can be made	Cause-and-effect conclusions can NOT be made	

these are randomly assigned to wear a nicotine patch; the other half are given a patch that has no nicotine. Every participant wears a patch for six weeks. Two months after the patch is removed (eight weeks after the start of the study), each study participant is assessed to determine whether or not he or she is smoking. Because study participants are volunteers and all volunteers meeting the study criteria were included, there was no random selection of units (people) for inclusion in the study. This would correspond to the second row in the body of Table 2.1.

Because there was no random selection of units, inference can be made *only* to the people included in the study. In practice, the argument is often made that the study participants are no different from other heavy smokers, and conclusions are made more broadly. However, if someone claims that these study participants are not representative of the population of heavy smokers and thus the conclusions do not apply to the whole population, there is no solid foundation by which to refute the claim.

The treatments (nicotine patch and no nicotine patch) were assigned at random. This corresponds to the second column in the table, so cause-and-effect conclusions can be made. That is, if the proportion of study participants who stopped smoking is significantly greater for those who wore the nicotine patch than the proportion of those who did not wear the nicotine patch, we conclude that the difference is due to the nicotine patch. The patch without nicotine is called a *placebo* patch because it has no active ingredients. Often, people who receive a treatment respond whether or not the treatment has any active ingredient. To be sure that a treatment, such as a patch or a pill, is effective, a treatment that appears the same but has no active ingredient is also given. The patch or pill or other item with no active ingredient is called a *placebo*.

In summary, the nicotine patch study is an experiment with a narrow scope of inference. It is an experiment because treatments are assigned at random. The scope of inference is narrow because people were not randomly selected from the population of heavy smokers for inclusion in the study, and thus, inference can be made only to the people in the study.

The nicotine patch study was a *blinded* one because the study participants did not know whether or not they had a patch with the active ingredient. A *double-blinded study* is one in which neither the study participant nor the individual determining the value of the response variable knows which treatment each person has received. By blinding, any tendency to favor one treatment over the other can be eliminated.

▶ Practice

For each of the following studies, specify the following:
   a. the response variable of interest
   b. the population to which inference may be made
   c. the types of conclusions (cause and effect or associations) that can be drawn
   d. the type of study

3. A drug company advertises for people to participate in a study on cholesterol. Those who answer the ad and pass the initial screening are randomly assigned to use either a standard medication or one that has recently been developed. After one month, the cholesterol level of each participant is measured.
4. A cable television company sends a questionnaire to 100 randomly selected customers to determine whether they would be willing to pay more if a new set of channels was added to the standard package.

5. A kindergarten teacher wants to study whether or not children who have been in day care learn the alphabet more rapidly than those who stay at home. After teaching the alphabet for the first month of school, the teacher randomly shows the letters to each student and records the number of correct responses. He also determines which children were in day care for at least one year prior to the start of school.

6. A large dental school wants to compare two approaches to cleaning teeth to determine which one takes less time. Among the patients who have teeth cleanings scheduled during the next six months, 60 are randomly selected and asked to participate in the study. Half are randomly assigned to have their teeth cleaned using one of the methods; the other half have their teeth cleaned using the other method. The time required to clean each person's teeth is recorded.

▶ In Short

In this lesson, we have discussed the key steps in conducting a study. Every decision made during a study has an impact on the analysis and interpretation of the results. The strongest conclusions are from experiments where treatments have been applied at random, allowing cause-and-effect conclusions to be made. Otherwise, we are limited to discussing observed associations between variables. By drawing the study units at random from a population, we are able to draw inference beyond the units used in the study to the population from which the units were drawn.

Describing and Displaying Categorical Data

LESSON SUMMARY

In the first lesson, we learned that the population is the set of objects or individuals of interest, and that the sample is a subset of the population. When presented with a set of either population or sample values, we need to summarize them in some way if we are to gain insight into the information they provide. In this lesson, we will focus on the presentation of categorical data, both categorically and graphically.

▶ Population Distribution versus Sample Distribution

Suppose that a high school orchestra has 62 members, 35 females and 27 males. The population of interest is the set of all orchestra members. Data are available on the categorical random variable, gender. The collection of genders for the population represents the population distribution of genders. In this example, the population is finite because there are a finite number of units (orchestra members) in the population.

Suppose 15 of the orchestra members are randomly selected. The genders of these 15 represent the sample distribution of gender. A summary measure of a population is called a *parameter*; a summary measure of a sample distribution, which is a function of sample values and has no unknown parameters, is called a *statistic*. Frequencies and relative frequencies are important summary measures for categorical variables such as the gender of the orchestra members.

As the sample size increases, the sample distribution tends to be more like the population distribution as long as the units for the sample have been drawn at random from the population. This is comforting in that, intuitively, we expect for the sample to "be better" as the sample size increases. What we mean by "better" is not always clear. However, the fact that the sample distribution begins to tend toward the population distribution is one way in which we have done better. Other measures of "better" will be discussed in later lessons.

▶ Frequency and Relative Frequency

The nature of categorical data leads to counts of the numbers falling within each category: the numbers of females and males; the numbers of red, yellow, or blue items; and the numbers ordering pizza, hamburger, or chicken. Notice, if we have two categories, we have two counts; three categories, three counts; and so on.

The number of times a category appears in a data set is called *the frequency of that category*. The *relative frequency of a category* is the proportion of times that category occurs in the data set; that is,

relative frequency = frequency/number of observations in the data set

These frequencies or relative frequencies are best organized in tabular form. The table should display all possible categories and the frequencies or relative frequencies. The frequency distribution (or relative frequency distribution) for categorical data is the categories with their associated frequencies (or relative frequencies). It is important to remember that if we have all population values, we can find the population frequencies and population relative frequencies. Otherwise, if we have a sample, which is a subset of the

population values, then we can find the sample frequencies and sample relative frequencies. For the band members, we have all population values. The population frequency and relative frequency distributions for the genders of these band members are displayed in Table 3.1.

Table 3.1 Gender of band members

CATEGORY	FREQUENCY	RELATIVE FREQUENCY
Female	35	35/62 = 0.56
Male	27	27/62 = 0.44
Total	62	1.00

Example

In the orchestra of 62 members, 36 play a string instrument, 12 play a woodwind, ten play brass, and four play percussion. Provide a tabular display of the frequency and relative frequency distributions for the type of instruments for this orchestra.

Solution

Table 3.2 Instrument distribution in orchestra

CATEGORY	FREQUENCY	RELATIVE FREQUENCY
String	36	36/62 = 0.58
Woodwind	12	12/62 = 0.19
Brass	10	10/62 = 0.16
Percussion	4	4/62 = 0.06
Total	62	1.00

Note that, although the frequencies sum properly to the total number of orchestra members, the relative frequencies actually sum to 0.99, not to 1 as indicated. The reason for this is *round-off error*. When dividing the category frequency by the total number of members, the result was not always an exact two-decimal value. We rounded to two decimal places. It is better to use the 1.00 as the total ($\frac{62}{62} = 1$) rather than reflecting the rounding error in the total.

Example

The orchestra is going to a special awards banquet during which dinner will be served. The hosts of the banquet need to know in advance whether the orchestra members prefer steak, fish, or pasta as their main dish. Thirty-two members choose steak, 12 choose fish, and 18 choose pasta. Provide a tabular display of the frequency and relative frequency distributions of the orchestra members' main dish choices.

Solution

Table 3.3 Orchestra members' dinner choices

CATEGORY	FREQUENCY	RELATIVE FREQUENCY
Steak	32	0.52
Fish	12	0.19
Pasta	18	0.29
Total	62	1.00

▶ Practice

1. Each of 83 students in a university class was asked whether he or she says "soda," "pop," or "coke" when ordering a cola beverage. Thirty-one responded "soda," 27 said "pop," and 25 use "coke." Provide a tabular display of the frequency and relative frequency distributions of the choice of term for cola beverages for this class.

2. For the 2004–2005 flu season, there was a shortage of the flu vaccine. The U.S. Centers for Disease Control (CDC) recommended that the vaccine be reserved for individuals at least 65 years old, those 6 to 23 months old, and persons aged 2 to 64 years with conditions that increased their risk of influenza complications. Persons aged 50 to 64 and those in household contact with high-risk people were added to the list in late December 2004. The city council of a large city wanted to know whether or not the final flu vaccination rate was high for community residents. In February 2005, they commissioned a survey of 1,000 randomly chosen residents who were at least 18 years of age. Of the 1,000 surveyed, 267 had received the vaccine and 733 had not. Provide a tabular display of the frequency and relative frequency distributions of those who had and had not received the flu vaccine.

▶ Visual Display for Categorical Data

Pie Charts

Pie charts and bar charts are common graphical approaches to displaying data. The frequencies, relative frequencies, or percentages can be presented graphically using pie charts or bar charts. For each category of chart, percent = relative frequency × 100%.

To make a pie chart, first draw a circle to represent the entire data set. For each category, the "slice" size is the category's relative frequency times 360 (because there are 360° in a circle). Each slice should be labeled with the category name. The numerical value of the frequency, relative frequency, or percentage associated with each slice should also be shown on the graph. Percentages are presented most commonly in pie charts. As an example, consider the frequency distribution of the high school orchestra members' gender. Thirty-five or 56.5% were female, and 27 or 43.5% were male. For the pie chart, the slice for females is $360 \times \left(\dfrac{35}{62} \right) = 203.22°$ of the pie; the remaining is for the males. See Figure 3.1.

Example

For the relative frequency distribution of the types of instruments played by the high school orchestra members discussed earlier in the lesson, create a pie chart.

Solution

First, the size of each pie must be found. For example, for the woodwinds, the slice of the pie is $\left(\dfrac{36}{62} \right) \times 360 = 209°$. After performing this calculation for each instrument type, the following pie chart can be created (see Figure 3.2).

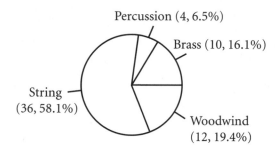

Figure 3.2

Example

Create a pie chart of the frequency and relative frequency distribution of the orchestra members' meal choice for the awards banquet.

Solution

See Figure 3.3 for an illustration of the orchestra members' meal choices.

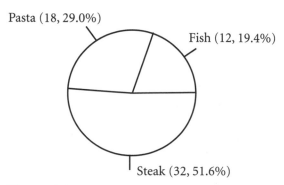

Figure 3.3

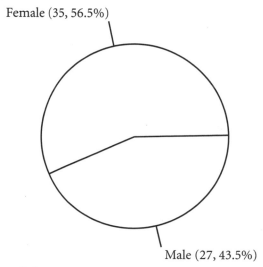

Figure 3.1

▶ Practice

3. Using the information in practice problem 1, create a pie chart of the frequency and relative frequency distributions for the use of "soda," "pop," and "coke" when referring to a cola beverage.

4. Using the information in practice problem 2, create a pie chart of the frequency and relative frequency distributions for flu vaccination or no flu vaccination during the 2004–2005 flu season based on this sample of the city's adult population.

Bar Charts

Bar charts may be used to display frequencies, relative frequencies, or percentages represented by each category in a data set. If there is only a response variable and frequencies are to be presented, a bar is used for each category and the height of the bar corresponds to the number of times that category occurs in the data set. For a relative frequency bar chart, a bar is used for each category, and the height of the bar corresponds to the proportion of times that response occurs in the data set. As an illustration, consider a relative frequency bar chart of the genders of the orchestra members.

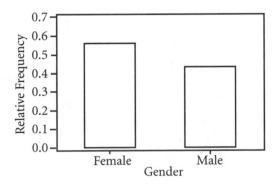

Figure 3.4

Notice that in Figure 3.4, both the x- and y-axes are labeled. Categories are displayed on the x-axis, and an appropriate scale is used on the y-axis.

Example

Create a frequency bar chart of the orchestra members' instruments.

Solution

See Figure 3.5 for a bar chart illustrating the instrument frequency.

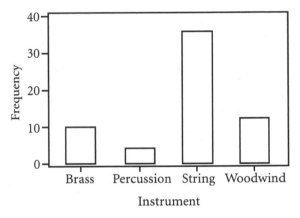

Figure 3.5

Example

Create a percent bar chart of the types of meals ordered for the awards banquet.

Solution

See Figure 3.6 for a bar chart of the percentages of meals ordered.

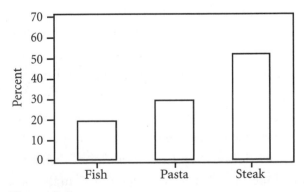

Figure 3.6

▶ Practice

5. For the university class discussed in practice problem 1, create a relative frequency bar chart of the responses that students gave in whether they called a cola beverage "soda," "pop," or "coke."

6. For the sample from the large city discussed in practice problem 2, create a percent bar chart of those 18 years and older who did and did not get the flu vaccination.

▶ Visual Displays for Categorical Data with Both Response and Explanatory Variables

Suppose a data set has a response and an explanatory variable, both of which are categorical. Pie charts and bar charts can be used to give visual displays of how the response variable might depend on the explanatory variable. For pie charts, a separate pie is created for each category of the explanatory variable.

Example

The instructor of the university class described in practice problem 1 was initially surprised by the results. Then she realized that whether the students used "pop," "soda," or "coke" in referring to a cola beverage probably depended on the region of the country in which the student grew up. She asked each student where he or she grew up *and* which term he or she used for a cola beverage. Of the students from the South, 25 used the term "coke" and four used "pop." Of the students from the Midwest, 17 used the term "pop" and 10 used "soda." Of the students from the Northeast, 21 used "soda" and six used "pop." No other region of the United States was represented in the class.

Present the data in tabular form showing both counts and relative frequencies for each response variable-explantory variable combination. Make side-by-side pie charts presenting the data.

Solution

The teacher has recorded the value of an explanatory variable (region) to help explain the response variable values (name used to refer to a cola beverage). It is important to notice that the relative frequencies of the response variable are computed for each region. As an illustration, the sum of the relative frequencies for the South will be one. The relative frequency of using the term "coke" for students from the South was $\frac{25}{29} = 0.86$. We do not divide by the total number in the class (83), only by the number from the South (29). See Table 3.4 and Figure 3.7 for illustrations of the data.

Table 3.4 Regions and term usage

REGION	TERM	COUNT	RELATIVE FREQUENCY
South	Coke	25	0.86
	Pop	4	0.14
Midwest	Pop	17	0.63
	Soda	10	0.37
Northeast	Pop	6	0.22
	Soda	21	0.78

Pie Chart of Midwest

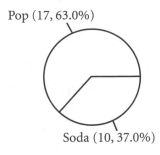

Pop (17, 63.0%)

Soda (10, 37.0%)

Pie Chart of Northeast

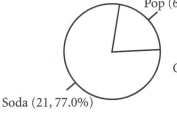

Pop (6, 22.2%)

Soda (21, 77.0%)

Pie Chart of South

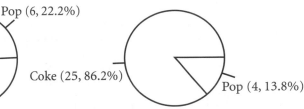

Coke (25, 86.2%)

Pop (4, 13.8%)

Figure 3.7

From the pie charts, it is easy to see that the terms used for cola beverages are quite different for various regions. Notice that these pie charts for the individual regions provide greater insight than the overall pie chart you created in practice problem 4.

▶ Practice

The City Council in the large city that did a study on the adults receiving flu vaccines decided it needed to consider the proportion receiving the vaccine within each risk group to judge the success of the program. Of the 233 sampled adults at least 65 years of age, 146 had been vaccinated. There were 158 sampled adults between ages 18 and 64 with high-risk conditions, and 41 of these were vaccinated. Healthcare workers with patient contact were another high-risk group, and 23 of the 64 sampled adults in this group received the flu vaccine. For those who did not fall in one of these high-risk groups, 24 of the 342 healthy sampled adults between 18 and 49 years of age and 33 of the 203 healthy sampled adults between 50 and 64 years were vaccinated.

7. List the response and explanatory variables.
8. Present the data in tabular form showing both counts and relative frequencies for each combination of the values of response and explanatory variables.
9. Make side-by-side pie charts presenting the data.

In a manner similar to the pie charts, bar graphs can be used to compare the response variable for varying values of the explanatory variable. Suppose we want to graph the relative frequencies. First, form a group of bars for each category of the explanatory variable. Within each group of bars, one bar is drawn for each category of the response variable. The height of the bar depends on the relative frequency.

Example
Create a bar chart that shows how the term used for a cola beverage changes with region of the country.

Solution

See Figure 3.8 for a bar chart showing cola terminology in different regions of the country.

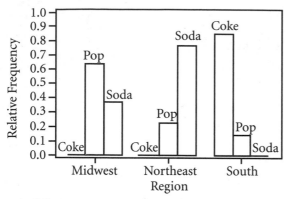

Figure 3.8

▶ Practice

10. For the complete flu vaccine data given in practice problem 7, create a bar chart that will allow for an easy comparison of the proportion vaccinated in each risk group.

▶ In Short

In this lesson, we have discussed the difference in population and sample distribution. As the sample size increases, the sample distribution tends to be more like the population distribution as long as sample units are selected at random. When working with categorical variables, pie charts and bar charts give a nice visual display of the data. Sometimes, an explanatory variable helps explain differences in responses. In these cases, multiple pie charts or side-by-side bar charts can help us visualize the relationship of the explanatory variable to the response.

Dotplots and Stem-and-Leaf Plots

LESSON SUMMARY

When numerical data are collected, we no longer have the counts for each category as we did for categorical data. Each observation is a number. We want to be able to display these numbers in a way that will provide more insight into the data. Displaying numerical data using dotplots and stem-and-leaf plots is the focus of this lesson.

▶ Dotplots

When numerical data are collected during a study, it is often difficult to understand what the numbers mean by simply looking at them. Recall the 62-member orchestra that we discussed in the beginning of Lesson 3. In addition to gender, the height (in inches) of each member was recorded. The heights and genders of the orchestra members are given in Table 4.1.

Table 4.1 Heights and genders of orchestra members

Gender	F	F	F	F	F	F	F	F	F	F
Height	57.5	60.5	70.2	62.8	69.1	71.9	65.2	59.7	66.6	65.0
Gender	F	F	F	F	F	F	F	F	F	F
Height	61.4	67.3	63.0	58.6	65.7	65.1	58.8	53.5	62.6	59.7
Gender	F	F	F	F	F	F	F	F	F	F
Height	64.3	59.7	64.8	59.3	68.1	60.5	65.8	66.4	62.5	67.1
Gender	F	F	F	F	F	M	M	M	M	M
Height	60.0	60.4	59.5	57.8	67.4	69.5	72.5	74.4	67.7	72.9
Gender	M	M	M	M	M	M	M	M	M	M
Height	66.3	67.8	68.9	67.4	64.9	67.7	83.8	72.6	71.5	75.0
Gender	M	M	M	M	M	M	M	M	M	M
Height	71.5	76.7	71.1	70.9	67.7	68.9	68.9	66.2	68.7	73.4
Gender	M	M								
Height	71.5	69.2								

Although we have all 62 heights of the orchestra members, it is difficult to describe the basic characteristics of the population distribution by looking only at the values. In this lesson, we will first look at some ways to display the data graphically. In the next lesson, we will consider some common measures used to summarize these population characteristics, beginning with measures of central tendency and measures of dispersion. Remember: The orchestra is the population of interest. Because we are summarizing information from the whole population, these measures are parameters. Later, we will consider how to estimate these values from a sample.

A *dotplot* is a simple way to present numerical data when the data set is reasonably small. To construct a dotplot, complete the following steps:

1. Draw a horizontal line and mark it with a measurement scale that extends at least as low as the smallest value in the data set and as high as the largest value in the data set.
2. For each observation in the data set, locate the value on the measurement scale and represent it by a dot. If two or more observations have the same value, stack the dots vertically.

A dotplot of the orchestra members' heights is shown in Figure 4.1.

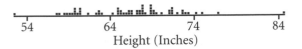

Figure 4.1

It is interesting to notice that we have *gaps* in the data. The member who is 53.5 inches tall is quite a bit shorter than the next shortest orchestra member. Similarly, the member who is 83.8 inches taller appears to be several inches taller than the next tallest orchestra member.

Because females tend to be shorter, on average, than males by the time they are in high school, it may be helpful to compare the distributions of female- and male-member heights. An effective way of doing this is to construct parallel dotplots. Here, parallel lines are drawn for each gender. The same scale is used for both lines, and the lines are labeled. Then dotplots are constructed for each group as in Figure 4.2.

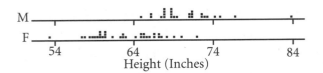

Figure 4.2

From the parallel dotplots shown in Figure 4.2, we see that, although some overlap exists, the males tend to be taller than the females. The shortest person is a female, and the tallest person is a male.

Example

A study was conducted to determine whether a person's "blinking rate" was, on average, different when playing video games than when engaged in conversation. Fifteen high school students were randomly selected from a large high school for participation in the study. The order in which treatments (video games and conversation) were applied to each individual was determined by the flip of a coin. The investigators were concerned that the study participants would, perhaps unconsciously, alter their blinking rate if they knew that was what was being observed. To avoid this, the study participants were told that swallow patterns were being studied and that the hypothesis was as follows: People will swallow less when playing a game or reading a book because they will produce less saliva due to the fight-or-flight reflex. Each participant was videotaped, and the number of blinks in a two-minute time interval was recorded for each treatment.

To illustrate how the study was conducted, suppose the first person to be tested was George. It was decided if the coin was heads-up, George would play the video games first; otherwise, the first treatment would be normal conversation. The coin was flipped, and the upper face was tails. George was engaged in normal conversation for about ten minutes. During this time, he was videotaped. The investigators wanted to pick a time period during the middle of the conversation to count the blink rate, so they counted the number of blinks in the fourth and fifth minutes of conversation. George was then given a 30-minute break. At the end of the break, he began playing a video game for ten minutes, during which time he was videotaped. The number of blinks in the fourth and fifth minutes was recorded. This process was repeated for each study participant. The results are shown in Table 4.2.

Table 4.2 Blinking frequency

STUDENT	NUMBER OF BLINKS DURING VIDEO GAME	NUMBER OF BLINKS DURING CONVERSATION
1	9	49
2	26	43
3	16	40
4	12	31
5	14	43
6	11	39
7	23	52
8	12	46
9	10	28
10	13	39
11	15	28
12	10	42
13	17	34
14	9	30
15	16	43

Based on what we learned in Lesson 2, this is an experiment with a broad scope of inference. It is an experiment because the treatments (that is, playing a video game and during normal conversation) are randomly assigned. In this case, the random assignment is the order in which the two are applied. The population is students in the large high school. Because the study participants were randomly selected from the population, conclusions can be made for all students in the high school; that is, a broad scope of inference exists. Can inference be drawn to all high school students? Perhaps, but that is not clear. If we assume that these high school students are representative of all high school students, then yes. However, these students may differ in some manner that would affect their blink rate while playing video games or during normal conversation. If so, then we could make erroneous conclusions if we tried to extend the results to all high school students.

The manner in which the study was conducted led to *paired data*. A pair of observations, one taken during normal conversation and the other taken while playing video games, is collected for each study participant. Because the objective of the study is to compare the two treatments, the difference in each pair of observations, representing the difference in the two treatments for that person, should be computed and these differences plotted. Most of us like to work with positive numbers more than negative numbers. Looking at the data, the number of blinks during normal conversation tends to be larger than the number of blinks while playing a video game, so we will compute the number of blinks during normal conversation minus the number of blinks while playing video games for each study participant.

Study Participant	1	2	3	4	5	6	7	8	9	10	11	12	13	14	15
Difference	40	17	24	19	29	34	18	26	13	26	13	32	17	21	27

Hint

When taking differences between treatments when data are paired, it is important to always take the difference in a consistent manner.

We could have taken the number of blinks while playing video games minus the number of blinks during normal conversation. Each difference would have had the same magnitude, but the opposite sign. Create a dotplot and discuss what the display implies about the data.

Solution

The dotplot of the 15 values of the differences in the number of blinks during normal conversation and video game playing is shown in Figure 4.3. Every participant blinked more during the two minutes of normal conversation than he or she did during the observed two minutes of playing the video games. (This may be seen because all differences are positive.) The smallest difference in the number of blinks was 13 and the largest was 40.

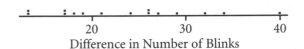

Difference in Number of Blinks

Figure 4.3

▶ Practice

1. Twenty high school students were randomly selected from a very large high school. They were asked to keep a record for a week of the number of hours they slept each night. These seven values were averaged to obtain an average night of sleep for each. The results are as follows: 9, 8, 8, 7.5, 6, 6, 4, 5.5, 7, 8, 5, 7.5, 6.5, 10, 8.5, 6.5, 5, 5.5, 7, and 7.5 hours. Create a dotplot of these data and discuss what the display implies about the data.

2. The ages of the Oscar winners for best actor and best actress (at the time they won the award) from 1996 to 2004 are as follows: 45, 39, 59, 33, 45, 25, 42, 24, 35, 32, 46, 32, 28, 34, 42, 27, 36, and 30. Create a doplot of these data and discuss what the display implies about the data.

3. Is there a difference in the ages of the Oscar winners who are best actors and best actresses? From 1996 to 2004, the ages of the best actors at the time of winning the award were 45, 59, 45, 42, 35, 46, 28, 42, and 36. The ages of the best actresses at the time of winning the award for this same time period were 39, 33, 25, 24, 32, 32, 34, 27, and 30. Create parallel dotplots for the actors' and actresses' ages. Comment on the similarities and differences that can be observed in the dotplots.

▶ Stem-and-Leaf Plots

A *stem-and-leaf plot* can be used to effectively display numerical data. Each number in the data set is broken into two pieces, a stem and a leaf. The stem is all but the last digit of the number, and the leaf is the last digit. To create a stem-and-leaf plot, do the following:

1. Determine the stem values. The stems should be equally spaced.

2. For each observation in the data set, attach a leaf to the appropriate stem. It is standard, though not mandatory, to put the leaves in increasing order at each stem value.

Look again at the heights of the 62 orchestra members. The stem-and-leaf plot of the data is shown as follows:

Stem								
53	5							
54								
55								
56								
57	5	8						
58	6	8						
59	3	5	7	7	7			
60	0	4	5	5				
61	4							
62	5	6	8					
63	0							
64	3	8	9					
65	0	1	2	7	8			
66	2	3	4	6				
67	1	3	4	4	7	7	7	8
68	1	7	9	9	9			
69	1	2	5					
70	2	9						
71	1	5	5	5	9			
72	5	5	9					
73	4							
74	4							
75	0							
76	7							
77								
78								
79								
80								
81								
82								
83	8							

As we saw in the dotplot, the orchestra members who are 53.5 and 83.8 inches tall have gaps between themselves and the next shortest and the next tallest members, respectively.

The stem-and-leaf plots of two groups can be compared using back-to-back stem-and-leaf plots. To construct one, a common stem is used. One group has leaves extending to the left of the stem; the other group's leaves extend to the right of the stem. Groups are labeled at the top. To compare the heights of female and male orchestra members, we would obtain the following back-to-back stem-and-leaf plot.

Females	Stem	Males
5	53	
	54	
	55	
	56	
8 5	57	
8 6	58	
7 7 7 5 3	59	
5 5 4 0	60	
4	61	
8 6 5	62	
0	63	
8 3	64	9
8 7 2 1 0	65	
6 4	66	2 3
4 3 1	67	4 7 7 7 8
1	68	7 9 9 9
1	69	2 5
2	70	9
9	71	1 5 5 5
	72	5 5 9
	73	4
	74	4
	75	0
	76	7
	77	
	78	
	79	
	80	
	81	
	82	
	83	8

The conclusions are similar to those from the parallel dotplots. The males tend to be taller than the females, though significant overlap exists in the heights of the two genders. One male was more than 7 inches taller than the next tallest orchestra member, and one female was more than 4 inches shorter than the next shortest orchestra member. In the next lesson, we will consider whether these values are unusual enough to be considered *outliers*.

Example

Draw a stem-and-leaf plot of the differences in the numbers of blinks during normal conversation and video game playing. Comment on the plot.

Solution

We might begin by using the tens column as the stem. This would result in the following graph:

1	3 3 7 7 8 9
2	1 4 6 6 7 9
3	2 4
4	0

Suppose now we decide to use seven stem values instead of four. The additional stems are created by making two stems for each of the 1, 2, and 3 stems. The first stem value is associated with the ones digits of 0 through 4; the second stem value is associated with the ones digits of 5 through 9. This stem-and-leaf plot is shown here:

1	3 3
1	7 7 8 9
2	1 4
2	6 6 7 9
3	2 4
3	
4	0

In the second plot, it is obvious that a gap exists in the data; no observations were made between 34 and 40. It appears 40 is an unusual value. With this one exception, in a two-minute period, the participants blinked between 13 and 34 more times during normal conversation than while playing a video game.

▶ Practice

4. Draw a stem-and-leaf plot for the average number of hours slept for the students in practice problem 1 and comment.

5. Draw a stem-and-leaf plot of the ages of the Oscar winners for best actor and best actress at the time they won, as given in practice problem 2, and comment.

6. Draw a back-to-back stem-and-leaf plot comparing the ages of the best actors to the ages of the best actresses for Oscar winners, based on the data in practice problem 3 and comment.

▶ In Short

Dotplots and stem-and-leaf plots are simple graphical presentations of the data. They provide a good view of the data. However, they are practical only for small- to medium-size data sets. We will look at other methods of presenting the data graphically in Lesson 7.

5 ▶ Measures of Central Tendency for Numerical Data

LESSON SUMMARY

In Lesson 4, we looked at dotplots and stem-and-leaf plots. Although we talked some about what we saw in the graphs, we need a greater statistical vocabulary if we are to describe these and other distributions. In this lesson, we will begin to think about measures of the middle of the distribution.

▶ Population Measures of Central Tendency

Measures of central tendency attempt to quantify the middle of the distribution. As we learned in Lesson 3, if we are working with the population, these measures are parameters. If we have a sample, the measures are statistics, which are estimates of the population parameters. There are many ways to measure the center of a distribution, and we will learn about the three most common: mean, median, and mode.

Mean

The *mean*, denoted by μ, is the most common measure of central tendency. The mean is the average of all population values. If a population has N members, the mean is

$$\mu = E(X) = \sum_{i=1}^{N} x_i = \frac{x_1 + x_2 + \ldots + x_N}{N},$$

where x_i is the value of the variable associated with the ith unit in the population. We have used two different notations to symbolize the mean. The first is the Greek letter μ, which has become a conventional representation of the mean. The other is $E(X)$, representing the "expected value of X." The average of a random variable across the whole population is the mean or the expected value of that variable. Notice that the symbol for a capital sigma (Σ) is a shorthand way of saying to add a set of numbers. The terms to be added are those beginning with $i = 1$ (because "$i = 1$" is below the sigma) to $i = N$ (because "N" is at the top of the sigma). The index, here i, is incremented by one in each subsequent term.

Example

Find the mean height of the 62 high school orchestra members given in Lesson 4.

Solution

The mean height for this population is

$$\mu = \frac{57.5 + 60.5 + \ldots + 69.2}{62} = 66.4.$$

Although we have presented a formal equation for the mean, it is important to remember that the population mean is simply the average of all population values.

Median

The *median,* another measure of central tendency, is the middle value of the population distribution. To find the median, order all of the values in the population from largest to smallest and find the middle value. For example, suppose the following five values constituted the population distribution:

$$2 \quad 3 \quad 7 \quad 8 \quad 22$$

The middle value is the 7. Now suppose the following four values represent the population distribution:

$$3 \quad 6 \quad 9 \quad 12$$

Here, the middle value is somewhere between the 6 and the 9. The median is any value between 6 and 9; however, usually, the average of the two values, $\frac{6 + 9}{2} = 7.5$, is taken as the median value. Through these two illustrations, we can see that we find the median in a slightly different manner when there is an even number of observations in the distribution than when there is an odd number of observations. This can be written generally as follows.

If N, the number of values in the population, is odd, the median is the $\frac{N + 1}{2}$st value in the ordered list of population values. If N is even, the median is any value between the $\left(\frac{N}{2}\right)$th and the $\left(\frac{N}{2} + 1\right)$st values in the ordered list of population values; usually, the average of the two values is taken as the median. Note that this definition of population median is appropriate only if the number of population units is finite. For a continuous random variable, the median is still the middle value in the population, but we must use other methods to define it.

If N, the number of values in the population, is odd, the median is the $\frac{N+1}{2}$st value in the ordered list of population values. If N is even, the median is any value between the $\left(\frac{N}{2}\right)$th and the $\left(\frac{N}{2}+1\right)$st values in the ordered list of population values; usually, the average of the two values is taken as the median.

Example

Find the median height of the 62 high school orchestra members. Compare the mean and median.

Solution

Because an even number of orchestra members exists, the median height is the average of the 31st and 32nd values in the ordered list of orchestra members' heights. The stem-and-leaf plot makes these values easy to find. Simply start at the top of the plot and count the number of leaves, always working from the stem out. Continue until the 31st and 32nd values have been identified. The 31st value is 66.6 inches, and the 32nd value is 67.1 inches. Any value between 66.6 and 67.1 is a median value. However, we will follow tradition and average the two: $\frac{66.6 + 67.1}{2} = 66.85$. That is, the median orchestra member height is 66.85 inches.

Notice that in this case, the mean and median are close, but not identical. For the median, exactly half of the population values are less than 66.85 inches, and half are greater than 66.85 inches. For the mean, 29 of the values are less than the mean, and 33 are greater than the mean, which is still close to half of the values. Sometimes, the mean and median are much further apart. We will consider what differences in the mean and median indicate about the distribution in the next lesson.

Mode

The *mode* is the most frequently occurring value in the population and is another measure of central tendency. This measure tends to be most useful for discrete random variables. For the orchestra members' heights, three members have heights of 59.7 inches. Similarly, we have three other groups, each with three members, with heights of 68.7, 68.9, and 70.5 inches. Thus, there are four modes.

Which measure of central tendency is the best? Each provides a little different information. The mean is the most common measure, but it is influenced by extreme values. One extreme value can have a big impact on the mean, especially if the population does not have many members. An unusually small population value may cause the mean to be quite a bit smaller than it would have been if that value was not in the population. Similarly, an unusually large population value may tend to inflate the mean. In contrast, the median and mode are not affected by these unusual values. The mode is often not useful because it may not be unique.

► Practice

As discussed in Lesson 4, from 1996 to 2004, the ages of the best actors at the time of winning the award were 45, 59, 45, 42, 35, 46, 28, 42, and 36. The ages of the best actresses at the time of winning the award for this same time period were 39, 33, 25, 24, 32, 32, 34, 27, and 30.

1. These are two populations, one for best actors and one for best actresses. Why are these populations and not samples?

2. Find the mean age of women winning the Oscar for best actress from 1996 to 2004.

3. Find the median age of men winning the Oscar for best actor from 1996 to 2004.

4. Find the median age of women winning the Oscar for best actress from 1996 to 2004.

5. Is the mode a useful measure of central tendency for either population? Explain.

6. Which of the measures of central tendency provides the best measure of the middle of these two population distributions? Explain.

7. Using the population parameters in the first five practice problems, discuss how the centers of these distributions compare.

► Sample Measures of Central Tendency

The population mean, median, and mode are parameters. To find their values, we must know *all* of the population values. Unless the population is small, as was the case when our population of interest was the orchestra members at one specific high school, this rarely happens. In practice, we cannot usually find the population mean, median, or mode. We estimate these parameters by finding the sample mean, sample median, and sample mode, respectively. The mean, median, and mode of the sample values are the *sample*

mean, *sample median*, and *sample mode*. These statistics are each an estimator of their population counterpart. That is, the *sample mean* is an estimate of the population mean; the *sample median* is an estimate of the population median; and the *sample mode* is an estimate of the population mode. We will learn more about the characteristics of these estimates and their uses later. For now, let's concentrate on how to find them.

Example

Find the sample mean, sample median, and sample mode for the difference in the number of blinks while playing video games and the number of blinks during normal conversation. Which one do you think is the best measure of central tendency for these data?

Solution

The sample mean is the average of the sample values, that is, the average of the sample differences:

$$\overline{X} = \frac{40 + 17 + \ldots + 27}{15} = 23.7$$

To find the median, we first order the sample values from smallest to largest:

13	13	17	17	18	19
21	24	26	26	27	29
32	34	40			

Because there are 15 values in the sample, the middle value is the $\left(\frac{15 + 1}{2}\right) =$ 8th value in the list; that is, the sample median is 24.

The sample mode is the most frequently occurring value in the sample. Here, the values of 13, 17, and 26 were each observed twice. In these cases, the sample mode is of little value in measuring central tendency.

The mean and median provide good measures of the center of distribution, but the mode does not.

▶ Practice

In Lesson 4, we created a dotplot and a stem-and-leaf plot for the average hours of sleep for 20 high school students who had been randomly selected from a very large high school. They had kept a record of the number of hours they slept each night for a week. These seven values were averaged to obtain an average night of sleep for each. The results were as follows: 9, 8, 8, 7.5, 6, 6, 4, 5.5, 7, 8, 5, 7.5, 6.5, 10, 8.5, 6.5, 5, 5.5, 7, and 7.5 hours.

8. These 20 data values represent a sample, not a population. What is the population from which they were drawn?

9. Find the sample mean.

10. Find the sample median.

11. Is the mode a useful measure of central tendency for these data?

12. Which of the measures of central tendency provides the best measure of center for these data? Explain.

▶ In Short

Mean, median, and mode are three measures of central tendency. Each provides a measure of the middle value in the population. The mean is the average of all population values. The median is the middle of the population values. The mode is the most frequently occurring population values. The sample mean, sample median, and sample mode are sample estimates of the corresponding population parameters.

Measures of Dispersion for Numerical Data

LESSON SUMMARY

As we discussed in Lesson 5, when describing a distribution, it is important to have some measure of where the middle is. With a little thought, we can see that, if we have a measure of the middle, we still need more information to describe the distribution fully. For example, are all of the population values the same or are they spread out over a range of values? If they are spread out, how should we measure the spread? Measuring the spread, or dispersion, of population or sample values is the focus of this lesson.

► Population Measures of Dispersion

Two distributions can have the same middle but still look very different. As an illustration, consider the two populations presented graphically in the dotplots in Figure 6.1.

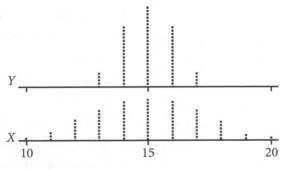

Figure 6.1

For each of the populations, the mean, median, and mode are all 15. Yet, they are different; one is more spread out than the other. The measures of dispersion are used to quantify the spread of the distribution. Range, interquartile range, mean absolute deviation, and standard deviation are four such measures that will be discussed in this lesson.

Range

The *range* is the difference in the largest and smallest population values; it is the total spread in the population. In Figure 6.1, Y assumes values from 13 to 17, giving a range of $17 - 13 = 4$. The range of X is $20 - 10 = 10$. Because the range of X is larger than that of Y, more than twice as large in this example, X is more spread out; that is, X has a larger dispersion than Y.

Example

Find the range of heights of the high school orchestra members that were first discussed in Lesson 4.

Solution

The tallest orchestra member is 83.8 inches tall, and the shortest is 53.5 inches tall. The range of the heights is $83.8 - 53.5 = 30.3$ inches, more than 2.5 feet!

▶ Practice

1. Find the range of the ages of the men who won the Oscar for best actor from 1996 to 2004. Their ages at the time of winning the award were 45, 59, 45, 42, 35, 46, 28, 42, and 36.
2. Find the range of the ages of the women who won the Oscar for best actress from 1996 to 2004. Their ages at the time of receiving the award were 39, 33, 25, 24, 32, 32, 34, 27, and 30.

Interquartile Range

Although the range in the heights of orchestra members is large (30.3 inches), most of the orchestra members are much closer in height than this indicates. The range is a crude measure of the dispersion of the population distribution. For example, consider the dotplot for the population distribution of Z in Figure 6.2.

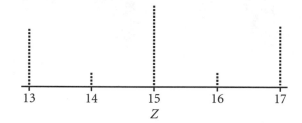

Figure 6.2

Notice the range of Z is 4 as is the range of Y in Figure 6.1. Both Y and Z have means, medians, and modes of 15. Yet, the distribution of Z appears to be more spread out than that of Y. We need another measure to capture this dispersion. We will explore two such measures, beginning with the interquartile range.

The range is greatly affected by exceptionally small or large values in a population. Instead of looking at the range of all population values, the *interquartile range* measures the spread in the middle half of the data. To find the interquartile range, we must first find the quartiles. The *quartiles* are the values that divide the population into fourths, just as the median divided the population in half. The first quartile separates the bottom 25% of the data from the top 75% of the data. The second quartile separates the bottom 50% of the data from the top 50% of the data. But, wait. That is exactly what the median does! The *second quartile* and the *median* are different names for the same quantity. The third quartile separates the bottom 75% of the data from the top 25% of the data.

The first and third quartiles are found by separating the lower half of the population values from the upper half of the population values. If there is an odd number of population values, the median is excluded from both halves. The first quartile, denoted by Q_1, is the median of the bottom half of the population. The third quartile, Q_3, is the median of the top half of the population values. The *interquartile range* (IQR) is then:

$$IQR = Q_3 - Q_1$$

The *IQR* is the range or spread of the middle half of the population values.

Example

Find the interquartile range of the population of orchestra members.

Solution

The median, 66.85 inches, divided the population into two halves, those members who are less than 66.85 inches tall and those who are greater than 66.85 inches tall. There are 31 members in each half. Because an odd number of values exists in the half, the median of the lower half is the $\left(\dfrac{31 + 1}{2}\right) = 16$th value in the ordered set of the 31 lower values. Thus, the first quartile is $Q_1 = 62.5$ inches. Similarly, the third quartile is the median of the 31 upper half values. The 16th value in the ordered set of upper values is $Q_3 = 69.5$ inches. The interquartile range is now found to be

$$Q_3 - Q_1 = 69.5 - 62.5 = 7 \text{ inches}$$

which is a value much smaller than the range. This is the spread in the middle half of the population of heights.

▶ Practice

3. Find the interquartile range of the ages of the men who won the Oscar for best actor from 1996

to 2004. Their ages at the time of winning the award were 45, 59, 45, 42, 35, 46, 28, 42, and 36.

4. Find the interquartile range of the ages of the women who won the Oscar for best actress from 1996 to 2004. Their ages at the time of receiving the award were 39, 33, 25, 24, 32, 32, 34, 27, and 30.

Mean Absolution Deviation

Before describing the next measure of dispersion, we need to define what is meant by a deviation. The quantity, $(X_i - \mu)$, is the deviation of the ith population value from the population. By taking the absolute value, the deviation is a measure of how far the value is from the mean. For the orchestra members, the shortest person has a deviation of $53.5 - 66.4 = -12.9$ inches. The negative sign indicates that this person's height is below the mean; that is, the shortest member's height is 12.9 inches below the mean height of all orchestra members. The member who is 74 inches tall has a deviation of $74 - 66.4 = 7.6$ inches. He is 7.6 taller than the average height of all orchestra members. If we add all of the deviations together, we get zero.

Mean absolute deviation is the mean (or average) distance of the population values from the population mean; that is,

$$\psi = E(|X_i - \mu|) = \frac{1}{N}\sum_{i=1}^{N}|X_i - \mu|.$$

Now, $(X_i - \mu)$ is the deviation of the ith population value from the mean. The absolute deviation of the ith population value from the population mean is the distance of that value from the population mean, $|X_i - \mu|$. For the populations of X, Y, and Z, we have $\mu = 15$. For a population value of 13, the distance of that value from the mean is $|13 - 15| = |-2| = 2$. The mean absolute deviation is the population mean of these distances, $E(|X - \mu|)$.

The mean absolute deviation of the population distribution of Y is 0.75, and that of Z is 1.14. Notice that the greater dispersion in the distribution of Z as compared to Y is captured in the mean absolute deviation.

Example

Find the mean absolute deviation of the heights of the 62 orchestra members. Interpret the mean absolute deviation in the context of the problem.

Solution

We begin by finding the deviation and the absolute deviation from the mean for each band member. The mean height was 66.4, so the deviation for a particular orchestra member is that member's height minus 66.4. The absolute deviation is the absolute value of the deviation, or how far the member's height is from the mean. These are given in Table 6.1.

The mean absolute deviation is the average of the absolute deviations. For the orchestra members, the mean absolute deviation is 4.32 inches. This means that, on average, an orchestra member's height is 4.32 inches from the population mean height of 66.4 inches.

Table 6.1 Height of orchestra members

Height	57.5	60.5	70.2	62.8	69.1	71.9	65.2	59.7	66.6
Deviation	−8.9	−5.0	3.8	−3.6	2.7	5.5	−1.2	−6.7	0.2
Absolute Deviation	8.9	5.0	3.8	3.6	2.7	5.5	1.2	6.7	0.2
Height	65.0	61.4	67.3	63.0	58.6	65.7	65.1	58.8	53.5
Deviation	−1.4	−5.0	0.9	−3.4	−7.8	−0.7	−1.3	−7.6	−12.9
Absolute Deviation	1.4	5.0	0.9	3.4	7.8	0.7	1.3	7.6	12.9
Height	62.6	59.7	64.3	59.7	64.8	59.3	68.1	60.5	65.8
Deviation	−3.8	−6.7	−2.1	−6.7	−1.6	−7.1	1.7	−5.9	−0.6
Absolute Deviation	3.8	6.7	2.1	6.7	1.6	7.1	1.7	5.9	0.6
Height	66.4	62.5	67.1	60.0	60.4	59.5	57.8	67.4	69.5
Deviation	0	3.9	0.7	−6.4	−6.0	−6.9	−8.6	3.0	3.1
Absolute Deviation	0	3.9	0.7	6.4	6.0	6.9	8.6	3.0	3.1
Height	72.5	74.4	67.7	72.9	66.3	67.8	68.9	67.4	64.9
Deviation	6.1	8.0	1.3	6.5	−0.1	1.4	2.5	1.0	−1.5
Absolute Deviation	6.1	8.0	1.3	6.5	0.1	1.4	2.5	1.0	−1.5
Height	67.7	83.8	72.6	71.5	75.0	71.5	76.7	71.1	70.9
Deviation	1.3	17.4	6.2	5.1	8.6	5.1	10.3	4.7	4.5
Absolute Deviation	1.3	17.4	6.2	5.1	8.6	5.1	10.3	4.7	4.5
Height	67.7	68.9	68.9	66.2	68.7	73.4	71.5	69.2	
Deviation	1.3	2.5	2.5	−0.2	2.3	7.0	5.1	2.8	
Absolute Deviation	1.3	2.5	2.5	0.2	2.3	7.0	5.1	2.8	

▶ Practice

5. Find the mean absolute deviation of the ages of the men winning the Oscar for best actor from 1996 to 2004. Their ages at the time of winning the award were 45, 59, 45, 42, 35, 46, 28, 42, and 36.

6. Find the mean absolute deviation of the ages of the women winning the Oscar for best actress from 1996 to 2004. Their ages at the time of receiving the award were 39, 33, 25, 24, 32, 32, 34, 27, and 30.

Variance and Standard Deviation

Two more measures of dispersion are the variance and the standard deviation. The *variance* is the mean squared distance of the population values from the mean; that is,

$$\sigma^2 = E\left[(X - \mu)^2\right] = \frac{1}{N}\sum_{i=1}^{N}(X_i - \mu)^2$$

Notice that we have expressed the variance as an expected value. Here, it represents the average squared deviation of a population value from the mean. For the population distribution of Y, displayed in Figure 6.1, the variance is

$$\sigma^2 = \frac{1}{100}\left[(13-15)^2 + (13-15)^2 + \ldots \right.$$
$$\left. + (17-15)^2\right] = 1.$$

For the population distribution of Z, displayed in Figure 6.2, the variance is 2.2. The units associated with the variance are the square of the measurement units of the population values. As an illustration, if the population values are recorded in inches, as they are

with the orchestra members' heights, the variance is in inches^2. To obtain a measure of dispersion in the same units as the population values, the *standard deviation* is found to be the square root of the variance; that is, $\sigma = \sqrt{\sigma^2}$.

Although not technically correct, the standard deviation is often described as being the average distance of a population value from the mean. For the population distribution of Y, the standard deviation is $\sqrt{1} = 1$. For Z, displayed in Figure 6.2, the population standard deviation is $\sqrt{2.2} = 1.5$. The mean absolute deviation and the standard deviation are the same for Y. This is very unusual. More commonly, the mean absolute deviation is close, but not equal to the standard deviation. For Z, the mean absolute deviation is 2, and the standard deviation is 1.5. Although these parameters are different, we will often interpret the standard deviation as we would the mean absolute deviation. We simply need to realize that this interpretation of the standard deviation is only an approximate one.

Which measure of dispersion should we use? It is always good to compute more than one measure of dispersion as each gives you slightly different information. The range is often reported. The mean absolute deviation is an intuitive measure of dispersion, but it is not used much in practice, primarily because it is difficult to answer more advanced statistical questions using mean absolute deviation. Most of the statistical methods are based on the standard deviation. However, the range, mean absolute deviation, and standard deviation are all inflated by unusually large or unusually small values in the population. In this case, the IQR is often the best measure of dispersion.

Example

Find the variance and standard deviation of the orchestra members' heights.

Solution

Carrying more significant digits than reported earlier, the mean height is 66.3774194 inches. The population variance is

$$\sigma^2 = \frac{1}{62}\left[(57.5 - 66.3774194)^2\right.$$
$$+ (60.5 - 66.3774194)^2 + \ldots +$$
$$\left.(69.2 - 66.3774194)^2\right] = 29.88;$$

that is, the variance of the orchestra members' heights is 29.88 inches^2. The standard deviation is $\sqrt{29.8807804} = 5.47$ inches.

▶ Practice

7. Find the variance and standard deviation of the men who won the Oscar for best actor from 1996 to 2004. Their ages at the time of winning the award were 45, 59, 45, 42, 35, 46, 28, 42, and 36.

8. Find the variance and standard deviation of the women who won the Oscar for best actress from 1996 to 2004. Their ages at the time of receiving the award were 39, 33, 25, 24, 32, 32, 34, 27, and 30.

9. Compare the range, interquartile range, mean absolute deviation, and standard deviation of the population of ages of men who won the Oscar for best actor from 1996 to 2004. Which measure do you prefer? Explain.

10. Compare the range, interquartile range, mean absolute deviation, and standard deviation of the population of ages of women who won the Oscar for best actor from 1996 to 2004. Which measure do you prefer? Explain.

▶ Sample Measures of Dispersion

We now have several measures of the spread of a population distribution. The challenge is that we generally have only a sample of values from the population, so we are unable to compute the population parameters. However, when we are interested in quantifying the spread of a distribution, we can use these sample values to estimate the population range, interquartile range, mean absolute deviation, variance, and standard deviation.

Sample Range and Sample Interquartile Range

The *sample range* is the largest sample value minus the smallest sample value. Often, the largest and/or the smallest population values are relatively rare in the population. If a small to moderately sized sample is drawn, then it is unlikely that both the largest and the smallest population values occur in the sample. Consequently, the sample range tends to underestimate the population range.

The difference in the first and third sample quartiles is the *sample interquartile range*. The process of finding the sample quartiles is the same as the one we used to find the population quartiles. First, find the median, or second quartile, of the sample values. The first quartile is the median of the bottom half of the data, and the third quartile is the median of the upper half of the data.

Sample Mean Absolute Deviation

The sample mean absolute deviation, $\widehat{\psi}$, is

$$\widehat{\psi} = \frac{1}{n}\sum_{i=1}^{n}|X_i - \overline{X}|.$$

where $\overline{X}$ is the sample mean and n is the sample size. Notice that the procedure for finding the sample mean absolute deviation is much like finding the population median absolute deviation. The differences are that the sample mean is used instead of the population mean, and only sample values are considered.

Sample Variance and Sample Standard Deviation

The sample variance, s^2, is computed as

$$s^2 = \frac{1}{n-1} \sum_{i=1}^{n} (X - \overline{X})^2.$$

Intuitively, we would like to use n instead of $(n-1)$ in this equation. However, if n instead of $(n-1)$ is used, the sample variance is a little smaller, on average, than the population variance; that is, the estimate of the variance would be *biased*. Although the previous equation allows us to see how the sample variance relates to the population value, it is somewhat cumbersome to compute it. A more computationally friendly way to find the sample variance is

$$s^2 = \frac{1}{n-1} \left[\sum_{i=1}^{n} X_i^2 - \frac{\left(\sum_{i=1}^{n} X_i \right)^2}{n} \right].$$

Look carefully at the previous equation. In computing $\sum_{i=1}^{n} X_i^2$, the sample values are squared and then added. For $\left(\sum_{i=1}^{n} X_i \right)^2$, the sample values are added together and then squared. The variance and sample variance are always greater than or equal to zero, so if you compute one of these and get a negative number, you know that you have made an error and should go back and redo the computations. Finally, the sample standard deviation is the square root of the variance; that is, $s = \sqrt{s^2}$.

Example

For the blink data, find the sample range, the sample interquartile range, the sample median absolute deviation, the sample variance, and the sample standard deviation.

Solution

The largest value in the sample is 40 and the smallest is 13, so the sample range is $40 - 13 = 27$.

There are 15 differences in the blink data set. The eighth value, 24, was found earlier to be the median. Because an odd number of sample values exists, the median is a sample value, and this value is excluded when finding the first and third quartiles. Seven values are below the median:

13	13	17	17	18	19	21

The median of these values is the fourth value, 17. This is the first sample quartile. There are also seven values above the median:

26	26	27	29	32	34	40

The median of these values is 29 and is the third sample quartile. The sample $I\widehat{Q}R$ is then $I\widehat{Q}R = 29 - 17 = 12$.

The sample mean was found to be 19.7. Thus, the sample mean absolute deviation is

$$\widehat{\psi} = \frac{1}{15} \left[(|40 - 19.6666667|, |17 - 19.6666667|, \right.$$
$$\left. \dots |27 - 19.6666667|) \right] = 6.3.$$

Notice that we reported the sample mean as 19.7, but when it came to computing the sample mean absolute deviation, we used 19.66666667. Why? Because we had only 15 sample values reported as whole numbers, we can claim, at most, one decimal of accuracy in estimating the population mean when using the sample mean, so we reported 19.7 instead of 19.6666667 or another number with even more decimal places. However, when we use this value in subsequent computations, it is best to carry as many digits as possible. Otherwise, the round-off error becomes larger in each step and can cause significant error in the final value. This is true for all computations that will be done in this workbook.

The sample variance is computed as

$$s^2 = \frac{1}{15 - 1}\left[(40^2 + 17^2 + \ldots + 27^2)\right.$$
$$\left. - \frac{(40 + 17 + \ldots + 27)^2}{15}\right]$$
$$= \frac{1}{14}\left[9{,}320 - \frac{356^2}{15}\right]$$
$$= 62.2,$$

and the sample standard deviation is found to be

$$s = \sqrt{62.20952} = 7.9.$$

That is, the sample variance is 62.2 blinks^2, and the sample standard deviation is 7.9 blinks in two minutes.

▶ Practice

In Lesson 4, we created a dotplot and a stem-and-leaf plot for the average hours of sleep for 20 high school students who had been randomly selected from a very large high school. They kept a record of the number of hours they slept each night for a week. These seven values were averaged to obtain an average night of sleep for each. In Lesson 5, we estimated the parameters that provide measures of central tendency. The 20 students slept an average of 9, 8, 8, 7.5, 6, 6, 4, 5.5, 7, 8, 5, 7.5, 6.5, 10, 8.5, 6.5, 5, 5.5, 7, and 7.5 hours.

11. Find the sample range.
12. Find the sample interquartile range.
13. Find the sample mean absolute deviation.
14. Find the sample variance and standard deviation.
15. Which of the measures of dispersion do you think is most appropriate for these data? Explain.

▶ In Short

Range, interquartile range, mean absolute deviation, variance, and standard deviation are all measures of the population dispersion. The range is the difference in the largest and smallest population values. The interquartile range is the difference in the third and first quartiles of the population values. The mean absolute deviation is the average distance of the population value from the population mean. The variance is the average squared distance of a population value from the mean, and the standard deviation is the square root of the variance. The sample range, sample interquartile range, sample mean absolute deviation, sample variance, and sample standard deviation are sample estimates of the corresponding population parameters.

Histograms and Boxplots

LESSON SUMMARY

Numerical data may be discrete or continuous. In this lesson, we will discuss presenting information on the distributions of discrete and continuous random variables in tabular form. Then we will learn how to display numerical data using histograms and boxplots. Discrete data most frequently arise from counting. In these cases, each observation is a whole number; however, some discrete data are not comprised fully of whole numbers. In contrast, a continuous random variable can take on any value in one or more intervals on the number line. Because a discrete random variable may assume either a finite or countably infinite number of values and a continuous random variable can assume any of an uncountably infinite number of values, we sometimes have to present data arising from observing these two types of random variables differently.

▶ Tabular Displays of Discrete Distributions

In Lesson 3, we organized categorical data in tabular form. For each category, the frequency and/or relative frequency were presented. We could not compute the cumulative relative frequency when working with categorical data because the categories had no natural ordering. If the number of possible values of a discrete random variable is finite, its population distribution can be displayed in a table, much like we did for the categorical random variable. For each possible value of the discrete random variable, the frequency or relative frequency is presented. Because for discrete numerical data the categories have an order (e.g., one is less than two is less than three), we may also want to record the cumulative relative frequencies in the table. The *cumulative relative frequency* of i is the number of observations, with the value of i or less divided by the total number of observations. If we have a sample from a discrete distribution and not the whole population, we can display the sample distribution in tabular form. For each observed value, we would have the frequency, relative frequency, and perhaps the cumulative relative frequency.

Example

A researcher named John L. Hoogland studied the mating behavior of Gunnison's prairie dogs at Petrified Forest National Park in Arizona for seven years, from 1989 to 1995. In 1998, he wrote an article titled "Why do female Gunnison's prairie dogs copulate with more than one male?" in *Animal Behavior* (55:351–359). Each year, all adult and juvenile prairie dogs at the 14-hectare study site were captured and marked. Mating season begins in mid-March and ends in early April. However, a female usually accepts partners on only one day of the breeding season. The number of partners accepted by each female prairie dog during a breeding season was recorded. During this seven-year period, female prairie dogs number 87, 93, 61, 17, and 5 accepted one, two, three, four, and five partners, respectively.

1. What type of a study was this?
2. Present the sample distribution in tabular form.

Solution

1. All female prairie dogs that could be observed were observed in the study area, so there was no random selection of prairie dogs from some larger population. The number of partners could not be assigned at random, so there was no random assignment of treatments. Therefore, this is an observational study (see the table in Lesson 2).

2. The sample distribution is shown in Table 7.1.

Table 7.1 Prairie dogs and partners

NUMBER OF PARTNERS	RELATIVE FREQUENCY	CUMULATIVE RELATIVE FREQUENCY
1	87	0.33	0.33
2	93	0.35	0.68
3	61	0.23	0.92
4	17	0.06	0.98
5	5	0.02	1.00
Total	263	1.00	

From the table, we see that 87 of the 263 prairie dog females accepted one partner, but five prairie dog females accepted five partners. Further, 93 (or 35%) of the females accepted two partners. As in Lesson 3, we have a rounding error because the relative frequency column sums to 0.99 and not to 1.0. Again, we report the 1.0 and not 0.99. It is better to accurately report the relative frequencies than to force columns or rows in a table to total to an inaccurate value. From the cumulative relative frequency column, we see that 68% of the prairie dogs accepted one or two partners.

▶ Practice

Use the following information to answer practice problems 1 and 2. A shoe manufacturer selected a random sample of male students attending a large university. The smallest shoe that will comfortably fit each selected male was recorded. Only whole shoe sizes were considered; no half sizes were included. In this study, 2, 6, 14, 13, 11, and 4 students were found to have shoe sizes of 7, 8, 9, 10, 11, and 12, respectively.

1. What type of a study was this?
2. Present the sample distribution in tabular form.

Grouped Discrete Data

When working with discrete data, each possible count is a natural category for a frequency distribution. Sometimes, numerous possible counts exist or a few large or small values are far away from most of the data. In these cases, a frequency distribution with a very long list of possible values does not aid understanding of the data. By grouping the observed values to form class intervals, or simply classes, greater insight into the data is often gained.

As an example, New Zealand's Tourism Research Council conducts an annual survey of international visitors. The length of stay of people traveling to New Zealand from other countries is collected. The results from the 2004 survey are displayed in Table 7.2 (Source: Tourism Research Council of New Zealand website at www.trcnz.govt.nz/Surveys/International+Visitor+Survey/Data+and+Analysis/).

Table 7.2 Length of tourist stay in New Zealand

LENGTH OF STAY IN NEW ZEALAND	INTERNATIONAL VISITORS DURING 2004	RELATIVE FREQUENCY
Under 5 days	483,276	0.22
5 to 7 days	395,232	0.18
8 to 10 days	235,689	0.11
11 to 13 days	175,138	0.08
14 to 16 days	169,971	0.08
17 to 19 days	114,835	0.05
20 to 29 days	235,683	0.11
30 or more days	340,283	0.16
Total	2,150,106	1.00

Suppose we had listed each length of stay in days. We would have had more than 30 classes. By grouping, we have eight classes, and it is easy to see that 22% of the international visitors stayed less than five days and 16% stayed at least 30 days. However, we have lost some information. We do not know how many stayed for three days or how many stayed for more than 60 days.

Tabular Displays of Continuous Data

The difficulty in displaying numerical data in tables becomes more pronounced when working with continuous numerical data. Consider the heights of the orchestra members discussed in Lessons 3 and 4. There were 62 orchestra members who had 52 different heights! Forming class intervals allows us to display the frequencies within each class. The challenge is that no natural intervals exist, so we have to define our own. Because the shortest orchestra member was 53.5 inches and the tallest 76.7 inches, it seems natural to begin the classes at 53 inches and stop them at 77 inches. The question is how long should each interval be? If we form intervals of 2 inches beginning at 53 inches, we would have 16 classes: 53–55, 55–57, 57–59, 59–61, 61–63, 63–65, 65–67, 67–69, 69–71, 71–73, 73–75, 75–77, 77–79, 79–81, 81–83, 83–85. If we choose 4-inch intervals, we would have eight classes: 53–57, 57–61, 61–65, 65–69, 69–73, 73–77, 77–81, 81–85. For now, let's look at both of these possibilities and later decide which one would be best.

We have one more problem that must be addressed before we can complete the frequency table: What happens if an observation falls on the boundary? As an illustration, one orchestra member is 65.0 inches tall. Should that person be in the 63–65 or the 65–67 class interval? We will adopt the convention that the lower boundary but not the upper boundary is included in a class. For the 65.0-inch-tall orchestra member, this means that she will be counted in the 65–67 class interval.

For 2-inch intervals, we would have the frequency table shown here in Table 7.3.

Table 7.3 Orchestra member height frequency (2-inch class intervals)

CLASS INTERVAL	FREQUENCY	RELATIVE FREQUENCY	CUMULATIVE RELATIVE FREQUENCY
53 to 55	1	0.02	0.02
55 to 57	0	0.00	0.02
57 to 59	4	0.06	0.08
59 to 61	9	0.15	0.23
61 to 63	4	0.06	0.29
63 to 65	4	0.06	0.39
65 to 67	9	0.15	0.53
67 to 69	13	0.21	0.74
69 to 71	5	0.08	0.82
71 to 73	8	0.13	0.95
73 to 75	2	0.03	0.98
75 to 77	0	0.00	0.98
77 to 79	0	0.00	0.98
79 to 81	0	0.00	0.98
81 to 83	0	0.00	0.98
83 to 85	1	0.02	1.00
Total	62	1.00	

Using the 4-inch class intervals, we obtain the frequency distribution shown in Table 7.4.

Table 7.4 Orchestra member height frequency
(4-inch class intervals)

CLASS INTERVAL	FREQUENCY	RELATIVE FREQUENCY	CUMULATIVE RELATIVE FREQUENCY
53 to 57	1	0.02	0.02
57 to 61	13	0.21	0.23
61 to 65	8	0.13	0.39
65 to 69	22	0.35	0.74
69 to 73	13	0.21	0.95
73 to 77	2	0.03	0.98
77 to 81	0	0.00	0.98
81 to 85	1	0.02	1.00
Total	62	1.00	

Before deciding whether to use the 2-inch or 4-inch class intervals, we will learn how to construct histograms.

▶ Practice

3. A bank wants to improve its customer service. Before deciding to hire more workers, the manager decides to get some information on the waiting times customers currently experience. During a week, 50 customers were randomly selected, and their waiting times recorded. The data are as follows: 18.5, 9.1, 3.1, 6.2, 1.3, 0.5, 4.2, 5.2, 0.0, 10.8, 5.8, 1.8, 1.5, 1.9, 0.4, 3.5, 8.5, 11.1, 0.3, 1.2, 4.4, 3.8, 5.8, 1.9, 3.6, 2.5, 4.5, 5.8, 1.5, 0.7, 0.8, 0.1, 9.7, 2.6, 0.8, 1.2, 2.9, 3.0, 3.2, 2.8, 10.9, 0.1, 5.9, 1.4, 0.3, 5.5, 4.8, 0.9, 1.6, and 2.2. Construct a tabular display of the data.

▶ Histograms

First, suppose we have ungrouped discrete data as in the number of partners for the Gunnison's prairie dog example. Constructing a histogram requires the following steps:

1. On the horizontal axis, draw a scale, mark the possible values, and label the axis.
2. On the vertical axis, draw a scale, mark it with either frequencies or relative frequencies, and label the scale.
3. For each possible value, draw a rectangle centered at that value with a height determined by the corresponding frequency or relative frequency.

Example

Construct the relative frequency histogram for the number of partners for Gunnison's prairie dogs.

Solution

The relative frequency histogram is displayed in Figure 7.1.

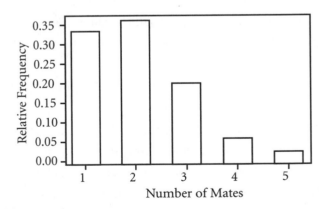

Figure 7.1

When working with continuous data, class intervals must be formed, as in Tables 7.3 and 7.4, before a histogram can be constructed. Once this is done, the process of constructing a histogram is similar to that for discrete data. For the 2-inch intervals of orchestra members' heights presented in Table 7.2, the histogram is shown in Figure 7.2.

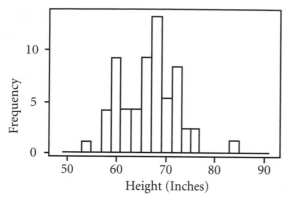

Figure 7.2

When looking at a histogram, you should look for a center value, the extent of spread or dispersion, the general shape, the location and number of peaks, and the presence of gaps and outliers. Here, perhaps the most notable feature is the three peaks at 60, 68, and 72 inches. These make determining the center, or typical, value a little difficult. The center of the data seems to be about 66 inches. This agrees well with the mean of 66.4 inches and the median of 66.85 inches found in Lesson 5. The spread seems to be from about 54 to 84 inches. An unusually small value and an unusually large value are separated from the rest of the observations by gaps.

In Figure 7.3, the class intervals are 4 inches wide (see Table 7.3). The center appears to be at about 67, still consistent with the mean and median found in Lesson 5. However, there are only two peaks now, one at about 59 inches and the other at about 67 inches. The shortest person no longer seems to be an outlier as no gap exists on the graph between that value and the next smallest value, but the tallest orchestra member continues to appear to be an unusual value.

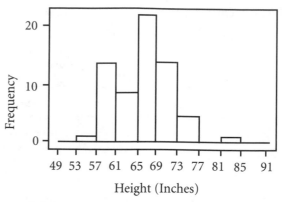

Figure 7.3

If class intervals are made too small, the average number of observations in each interval is small and subject to quite a bit of variation. This results in fluctuations in the height of the bars that may simply reflect fluctuations in the data and not true distributional characteristics. In contrast, if the class intervals are too wide, important features of the data may be obscured. The histogram using 2-inch intervals appears to reflect the fluctuations in the data, whereas the one based on 4-inch intervals is more clearly reflecting the distributional characteristics. Thus, 4-inch intervals are the most appropriate. As a general rule, the number of classes is often set approximately to the square root of the number of data points. Using this rule, seven or eight classes would be about the right number as there are 62 orchestra members. This agrees with our choice of 4-inch intervals.

▶ Practice

4. Create a relative frequency histogram for the number of partners accepted by female Gunnison's prairie dogs in the study area at the Petrified Forest National Park during the seven-year study period presented earlier in the lesson.

5. Create a relative frequency histogram of the number of days international visitors stayed in

New Zealand during 2004 based on the data given earlier in the lesson.

6. Create a relative frequency histogram of the waiting times of bank customers based on the data given earlier in the lesson.

▶ Boxplots

Another graph that is extremely useful is the boxplot. To create a boxplot, we need the five-number summary. The *five-number summary* consists of the first quartile, the median, the third quartile, the smallest observed value, and the largest observed value. The following steps lead to a boxplot:

1. Draw a scale that extends below the smallest and above the largest values in the data set on either the horizontal or vertical axis.

2. Draw parallel line segments at the first quartile, the median, and the third quartile. Connect the ends of the three parallel line segments to form a box.

3. Extend "whiskers" from the center of the first quartile line segment and the center of the third quartile line segment to the smallest and largest observations, respectively, as long as these most extreme observations are within 1.5 IQR of the closest quartile. Otherwise, extend them to the smallest value within 1.5 IQR of the first quartile and to the largest value within 1.5 IQR of the first quartile.

4. If there are any observations beyond 1.5 IQR of the nearest quartiles (so the whiskers do not extend to these values), mark these observations with an asterisk (*).

5. The mean may also be marked using a circle, which allows an easy comparison of the relative sizes of the mean and median.

Any value that is more than 1.5 IQR units below Q_1 or above Q_3 is defined to be an outlier.

Example

Create a boxplot for the orchestra members' heights introduced in Lesson 4. Identify any outliers that might be present.

Solution

Referring again to our work with the orchestra members' heights in Lessons 5 and 6, we have the following five-number summary:

First quartile (Q_1):	62.5
Median (Q_2):	66.85
Third quartile (Q_3):	69.5
Smallest value:	53.5
Largest value:	83.8

As determined in Lesson 6, the IQR for the orchestra members' heights is 7 inches; thus, 1.5 IQR = 1.5 (7) = 10.5 inches. $Q_1 - 1.5\ IQR = 62.5 - 10.5$ = 52.5 inches. Because the smallest orchestra member is 53.5 inches, which is greater than the 52.5 just calculated, the lower whisker extends only to 53.5 inches. Now, $Q_3 + 1.5\ IQR = 69.5 + 10.5 = 80$ inches. The two tallest orchestra members are 76.7 and 83.8 inches tall. The member who is 76.7 inches tall is within 1.5 IQR of the third quartile, but the tallest member is not. Therefore, the upper whisker extends from Q_3 to 76.7 (the largest observation within 1.5 IQR of Q_3), and a star is used to designate the 83.8 inches that is beyond the end of the whisker. Finally, the mean is denoted with a circle in the box.

Here, the orchestra member who is 83.8 inches tall is an outlier, but the shortest member who is 53.5 inches tall is not. Notice that the histogram based on 4-inch class intervals reflects which values are outliers better than the one based on 2-inch class intervals (see Figure 7.4).

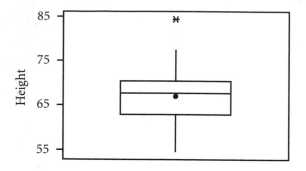

Figure 7.4

▶ Shape of a Distribution

Unlike dotplots and stem-and-leaf plots, histograms and boxplots may be used with very large data sets. All four, but especially the histograms and boxplots, provide a visual display of the shape of the distribution. Three specific shapes will be discussed most frequently: symmetric, right skewed, and left skewed. If a vertical line can be drawn through the center of a histogram such that the area to the left of the line is a mirror image of the area to the right, the distribution is *symmetric* (see Figure 7.5). For boxplots, a distribution is symmetric if the shape of the box and length of the whiskers for observations that are smaller than the

median is a mirror image of the shape of the box and length of the whiskers for observations that are greater than the median. The most common continuous distribution that is unimodal and symmetric is the normal distribution, which we will discuss in Lesson 11.

If a distribution has one mode (is unimodal) and is not symmetric, the distribution is said to be *skewed*. Proceeding to the right of the mode in a unimodal distribution, we move to the upper, or right, tail of the distribution. Similarly, we move to the lower, or left, tail of the distribution as we proceed to the left of the mode in a unimodal distribution. If the upper tail stretches out farther than the lower tail, the distribution is right or positively skewed (see Figure 7.5). If the lower tail stretches out farther than the upper tail, the distribution is left or negatively skewed (see Figure 7.5). These shapes may be seen in both histograms and boxplots.

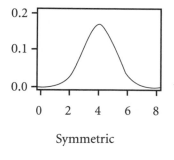

Symmetric

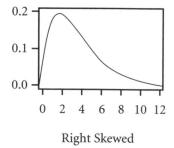

Right Skewed

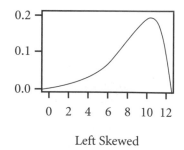

Left Skewed

Figure 7.5

▶ Practice

7. Create a boxplot of the number of partners accepted by female Gunnison's prairie dogs in the study area of the Pertrified Forest National Park during the 7-year study period presented in Table 7.1.

8. Create a boxplot of the waiting times of bank customers based on the data given in practice problem 3 earlier in this lesson.

▶ In Short

Numerical data can be summarized in tabular form. However, if the number of possible values of a discrete random variable is large or if the random variable is continuous, then possible values need to be grouped. Frequency or relative frequency histograms and boxplots provide visual summaries of the distributions.

Describing and Displaying Bivariate Data

LESSON SUMMARY

In the first seven lessons, we have considered only one numerical measurement from each observational unit. Often, two or more numerical measurements are taken on each observational unit. For example, the weight and length of each fish taken from a lake may be recorded. In studies of wetlands, the phosphorus and nitrogen concentrations at several sites might be observed. In another study, the goal was to determine whether a person's ear grows throughout life. For each person in the sample, age and the length of the left ear were recorded. In these studies, we are interested in not only the values of each of the variables assumed, but also in how they relate to each other. In this lesson, graphical displays for bivariate data and a measure of the relationship of two variables will be explored.

▶ Scatter Plots

Suppose that more than one, say two, numerical values are recorded for each unit in the study. Sometimes, both of the variables are responses. At other times, we are interested in how the *response variable*, or *dependent variable*, relates to the *explanatory variable*, or *independent* or *predictor variable*. In the latter case, we may want to predict the value of the response variable for a specific value of the explanatory variable. More than one explanatory variable may exist for each response variable.

When working with univariate data, we saw that plots (pie charts, bar charts, dotplots, stem-and-leaf plots, histograms, and boxplots) aided our understanding of the data. Although we could construct these types of graphs for each variable and gain a better understanding of each variable, such graphs would not aid our understanding of how the two variables are related. A scatter plot is an effective graph for gaining insight into bivariate data. A *scatter plot* is a graph in which each observation (pair of numbers) is represented by a point in a rectangular coordinate system. The horizontal axis is identified with the x-axis, and the axis is scaled to cover the range of values of X. The vertical axis is identified with the y-axis, and the axis is scaled to cover the range of values of Y. If both an explanatory and a response variable exist, the x-axis is used for the explanatory variable, and the y-axis is used for the response variable. The point representing the observation (x,y) is placed at the intersection of the vertical line through x on the x-axis and the horizontal line through y on the y-axis. Figure 8.1 shows the point representing the observation (2.5,3) with the corresponding vertical and horizontal lines. These reference lines are not usually included in the plot; they are here only for illustration. All data points would be plotted using the same approach.

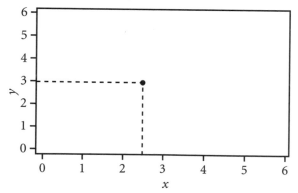

Figure 8.1

Example

A group of students wanted to know whether there was a relationship in the height from which a ball was dropped and its rebound height. Using a basketball, they dropped the ball from each of 11 heights three times and measured how high it rebounded. Both the height from which the ball was dropped and the height of the rebound were measured, in inches, from the bottom of the ball. The data are given in Table 8.1.

Table 8.1 Ball drop height and rebound height

DROP HEIGHT	REBOUND MEASUREMENT 1	REBOUND MEASUREMENT 2	REBOUND MEASUREMENT 3
12	3	6	5
18	7	8	11
24	13	14	16
30	19	18	17
36	20	21	20.5
42	22	21.5	21
48	24	26	25
54	27	28	30
60	25	31	32
66	37	39	38
72	45	44	42

1. Specify whether drop height and rebound height are explanatory or response variables.
2. Create a scatter plot of the data and describe the relationship between drop height and rebound height.

Solution

Drop height is the explanatory variable because this is the variable that is controlled during the study. Rebound height is the response variable because the rebound height was measured for a given drop height. Thus, drop height is on the *x*-axis and rebound height is on the *y*-axis. The data are plotted in Figure 8.2.

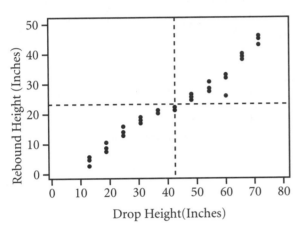

Figure 8.2

The ball did not always have the same rebound height when it was dropped repeatedly from a specific drop height; there was variability in the rebound heights for a given drop height. The rebound height tends to increase in a linear manner as the drop height increases, though the relationship is certainly not an exact one.

▶ Practice

A researcher thought that the average high temperature in March for U.S. cities could be explained by latitude. He collected data on the average high in March and the latitudes for 16 U.S. cities. The data are listed in Table 8.2.

1. We have two variables: temperature and latitude. Specify which variable is the explanatory variable and which is the response variable.

2. Create a scatter plot of the data and describe the relationship between latitude and average temperature in March.

Table 8.2 Temperature of U.S. cities

TEMPERATURE (°F)	LATITUDE (N)
34	61
50	40
75	25
60	37
56	45
69	32.7
41	44.9
41	44.3
46	42.2
47	41.8
55	38.6
82	27.5
69	34
53	47.4
79	24.4
79	21.3

▶ Pearson's Correlation Coefficient

One of the challenges in working with two or more variables is that they could have different units of measurements (inches, pounds, liters, etc.), means, standard deviations, or other characteristics. It is often

Pearson's correlation coefficient is defined to be:

$$r = \frac{1}{n-1} \sum_{i=1}^{n} z_X z_Y$$

helpful to have all variables on a common scale. Although there are many possible scales, we transform the original values of each variable so that the mean is zero and the standard deviation is one. *z-scores* are the transformed values of a random variable that have a mean of zero and a standard deviation of one; that is,

$$z = \frac{observation - mean}{standard\ deviation}.$$

If all population values are known, the population mean and standard deviation are used to find the z-score; if sample values are available, the sample mean and sample standard deviation are used to find the z-scores.

Let $(x_1, y_1), (x_2, y_2), \ldots, (x_n, y_n)$ be a random sample of n (x,y) pairs. Suppose we replace each x-value by its z-score, z_X, by subtracting the sample mean, $\bar{x}$, and dividing by the sample standard deviation, s_X. (Note that the subscript on s indicates the variable, here X, for which s is the sample standard deviation. Subscripts are often used in this manner when more than one variable is of interest in order to avoid confusion.) Similarly, suppose that each y-value is replaced by its z-score, z_Y. Note that, if x (or y) is larger than the sample mean $\bar{x}$ (or $\bar{y}$), z_X (or z_Y) is positive. Likewise, if x (or y) is smaller than the sample mean $\bar{x}$ (or $\bar{y}$), z_X (or z_Y) is negative.

Consider the sample of (x,y) pairs displayed in the graph in Figure 8.3. It is clear that there is a strong positive relationship between X and Y. The dashed horizontal line through $\bar{x}$ and the dashed vertical line through $\bar{y}$ divide the graph into four quadrants, which are labeled I, II, III, and IV. In quadrant I, both x and

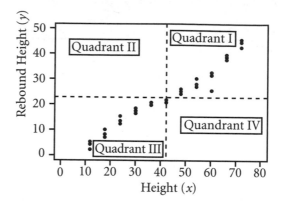

Figure 8.3

y are above their respective sample means; thus, z_X and z_Y are positive, and $z_X z_Y$ is positive. For (x,y) in quadrant II, x is below its sample mean and y is above its sample mean; therefore, $z_X z_Y$ is negative. Notice that $z_X z_Y$ is positive in quadrant III because z_X and z_Y are both negative and the product of two negative numbers is a positive number. Finally, because x is above its mean and y is below its mean, $z_X z_Y$ is negative in quadrant IV. Notice that, for the rebounding ball example, almost all of the points are in quadrants I and III, so $\sum_{i=1}^{n} z_X z_Y$ would be positive. In contrast, if most of the points lie in quadrants II and IV, $\sum_{i=1}^{n} z_X z_Y$ would be negative.

These ideas are the foundation for Pearson's correlation coefficient r, which provides a measure of the strength of the linear relationship between X and Y. Pearson's correlation coefficient is defined to be

$$r = \frac{1}{n-1} \sum_{i=1}^{n} z_X z_Y.$$

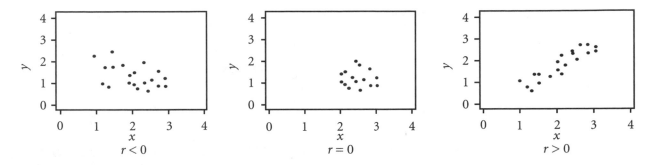

Figure 8.4

The correlation coefficient has some important properties. First, the value r is unitless; that is, it does not depend on the unit of measurement of either variable. X and Y can be measured in inches, meters, or light years, and the value of r would not change. Second, it does not matter which variable is labeled X and which is labeled Y; the value of r will be the same. Third, Pearson's r is always between -1 and $+1$. A value of one or -1 occurs when an exact linear relationship exists between X and Y. If $r = 1$, the slope of the line is positive; if $r = -1$, the slope of the line is negative. The closer r is to 1 or -1, the stronger the linear relationship between X and Y is. Finally, it is important to realize that r measures only the linear relationship in X and Y. It is possible for X and Y to have a very strong relationship and for r to be near zero. In these cases, the strong relationship is not linear in nature. Some scatter plots with the associated r values are shown in Figure 8.4.

Example

1. Find the z-scores associated with the drop heights and the rebound heights of the dropped basketball.
2. Find the Pearson's correlation coefficient for these data.
3. Relate the correlation coefficient r to the graph.

Solution

1. Many calculators have a built-in function that can be used to compute the correlation coefficient. We will not use that function here, but instead demonstrate a way to organize the computations required to find r when such a function key is not available. First, we need to find the sample mean and sample deviation for the drop height and for the rebound height. The sample mean for the drop height is 42 inches, and the sample standard deviation of drop height is 19.2678489 inches. The sample mean for the rebound height is 22.9090909 inches, and the sample standard deviation is 11.4390057 inches. Notice that we are carrying all of the decimal places at this point to reduce the effect of a rounding error on our value for r. If we were to report these values, we would round to about one decimal place. Using the sample means and sample standard deviations, we find the z-scores for drop height and rebound height for each observation as well as the product of the two. These are presented in Table 8.3.

   The values of z_{Drop} and $z_{Rebound}$ sum to zero because the transformation of the sample values was designed to give a mean of zero, making the total also zero.

2. Pearson's r is found using the total of the product of these two values; that is,

$$r = \frac{1}{n-1} \sum_{1}^{n} z_{Drop} z_{Rebound} = \frac{1}{32}(31.2379436) = 0.976.$$

Table 8.3 z-scores for drop height and rebound height

DROP HEIGHT	REBOUND HEIGHT	z_{Drop}	$z_{Rebound}$	$z_{Drop}z_{Rebound}$	DROP HEIGHT	REBOUND HEIGHT	z_{Drop}	$z_{Rebound}$	$z_{Drop}z_{Rebound}$
12	3.0	−1.55700	−1.74046	2.70989	42	21.0	0.00000	−0.16689	0.00000
12	6.0	−1.55700	−1.47820	2.30155	48	24.0	0.31140	0.09537	0.02970
12	5.0	−1.55700	−1.56562	2.43766	48	26.0	0.31140	0.27021	0.08414
18	7.0	−1.24560	−1.39078	1.73235	48	25.0	0.31140	0.18279	0.05692
18	8.0	−1.24560	−1.30336	1.62346	54	27.0	0.62280	0.35763	0.22273
18	11.0	−1.24560	−1.04109	1.29679	54	28.0	0.62280	0.44505	0.27718
24	13.0	−0.93420	−0.86625	0.80925	54	30.0	0.62280	0.61989	0.38607
24	14.0	−0.93420	−0.77883	0.72759	60	25.0	0.93420	0.18279	0.17076
24	16.0	−0.93420	−0.60399	0.56425	60	31.0	0.93420	0.70731	0.66077
30	19.0	−0.62280	−0.34173	0.21283	60	32.0	0.93420	0.79473	0.74243
30	18.0	−0.62280	−0.42915	0.26728	66	37.0	1.24560	1.23183	1.53437
30	17.0	−0.62280	−0.51657	0.32172	66	39.0	1.24560	1.40667	1.75215
36	20.0	−0.31140	−0.25431	0.07919	66	38.0	1.24560	1.31925	1.64326
36	21.0	−0.31140	−0.16689	0.05197	72	45.0	1.55700	1.93119	3.00686
36	20.5	−0.31140	−0.21060	0.06558	72	44.0	1.55700	1.84377	2.87075
42	22.0	0.00000	−0.07947	0.00000	72	42.0	1.55700	1.66893	2.59852
42	21.5	0.00000	−0.12318	0.00000	Total		0	0	31.2379436

3. This high value of r, with the scatter plot of the data, allows us to conclude that a line would provide a good model for these data, at least in the range of drop heights considered in this study. It is important to look both at the scatter plot and r, not just r by itself, before drawing a conclusion of a linear relationship as the following example demonstrates.

Because we have a sample for the rebound heights, Pearson's r is an estimate of the population correlation coefficient rho, or ρ. The population correlation coefficient has the same basic properties as r. However, it is important to remember that ρ does not change; it is a population characteristic. In contrast, r is based on a sample. If another sample is drawn from the same population, the value of r is very likely to change. Each sample provides an estimate of *rho*. In the example of dropping a basketball, we have a sample of size $n = 33$. We can use the data to estimate the population correlation. If we were to conduct the study again, we would undoubtedly obtain a similar but different value of r, which would also be an estimate of the population correlation.

Example

We have collected the following (x,y) pairs: $(-8.7, -0.6), (-8.2, -1.3), (-6.1, -2.0), (-4.1, -4.0),$ $(-1.6, -5.5),$ $(-0.2, -6.0),$ $(0.7, 4.6),$ $(1.4, 4.2),$ $(3.8, 3.8), (6.5, 3.1), (8.2, 2.0),$ and $(9.1, 0.6).$

1. Construct a scatter plot of the data and discuss the relationship in X and Y based on the plot.
2. Find Pearson's correlation coefficient r.
3. Discuss the relationship in r and what was observed in the graph.

Solution

The scatter plot of the data is shown in Figure 8.5.

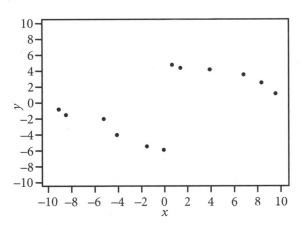

Figure 8.5

It appears that there are two groups of responses. The group in the upper right portion of the graph seems to have decreasing y-values as x increases. Similarly, the group in the lower left portion of the graph seems to have decreasing y-values as x increases.

To find Pearson's correlation coefficient, we organize the data in tabular form; find z_X, z_Y, and $z_X z_Y$, and the columns totaled (see Table 8.4).

As expected, the sums of z_X and z_Y are zero. The product of the two is used to find Pearson's r:

$$r = \frac{1}{n-1} \sum_{1}^{n} z_X z_Y = \frac{1}{11}(5.2422074) = 0.475$$

This value of r would tend to lead us to believe that there is a moderately strong, positive, linear relationship in X and Y. This would be the wrong conclusion for these data. This is why it is so important to look at the scatter plot when interpreting r. If we consider the two groups separately, r for the group with negative x values is -0.992, and r for the group with positive x values is -0.941. Both of these suggest a

Table 8.4

x	y	z_X	z_Y	$z_X z_Y$
−8.7	−0.6	−1.43825	−0.13524	0.19451
−8.2	−1.3	−1.35622	−0.32147	0.43599
−6.1	−2.0	−1.01170	−0.50771	0.51365
−4.1	−4.0	−0.68358	−1.03980	0.71079
−1.6	−5.5	−0.27343	−1.43887	0.39343
−0.2	−6.0	−0.04375	−1.57189	0.06877
0.7	4.6	0.10390	1.24820	0.12969
1.4	4.2	0.21875	1.14178	0.24976
3.8	3.8	0.61249	1.03537	0.63415
6.5	3.1	1.05545	0.84913	0.89622
8.2	2.0	1.33435	0.55648	0.74254
9.1	0.6	1.48200	0.18402	0.27271
Totals		0	0	5.2422074

strong, negative relationship in X and Y. When something like this happens, the researcher must try to determine what the difference is in the two groups. If the observations were taken from people, potential factors such as gender, age, and disease would represent the two groups.

▶ Practice

Look again at the data on the relationship of the average monthly temperature in March and latitude.

3. Find the z-scores associated with the average monthly temperature in March and the latitude.
4. Find the Pearson's correlation coefficient for these data.
5. Relate the correlation coefficient r to the graph.

▶ In Short

Bivariate data arise often in studies. The relationship in the two variables is often of great interest. Scatter plots are visual displays of the data that help us understand how the two variables might be related. Pearson's correlation coefficient r is a measure of the strength of the linear relationship in the variables. It is important to look at the scatter plot when interpreting the meaning of r.

LESSON

9 ▶ Basic Ideas in Probability

LESSON SUMMARY

In the first eight lessons, we have discussed different types of studies and how to summarize data both numerically and graphically. Yet, for most studies, we want to do more than this. We want to use the study to make decisions or to draw inferences about the research question. Before we are prepared to do this, we need to establish some understanding of probability, discrete and continuous distributions, and sampling distributions. These will be the topics of this and the next three chapters. We will begin by first introducing some probabilistic concepts.

▶ Sample Spaces and Events

People use probability to make decisions daily. The weatherman says that the chance of rain is 60% in the afternoon, and you decide it may be a good idea to take an umbrella. An economist predicts that the stock market will go up during the next six months, and you decide to buy some stocks. A friend took a class last semester that you are now taking. He says that he is sure you can pass without studying for any test, and you must decide whether or not to study. Some food in the refrigerator is past its expiration date, but it smells okay. Do you eat it or not? Based on our past experiences, we make these decisions, primarily without thinking about probability. Now, we will begin to look at probability more formally.

To begin, suppose we roll a fair die and observe the number of dots on the upper face. This is a *random experiment* because a chance process (rolling the die) determines the outcome. The set of all possible outcomes is called the *sample space* and will be denoted by S. For our die rolling experiment, the sample space is $S_1 = \{1, 2, 3, 4, 5, 6\}$.

Suppose that we collect fish randomly from a lake and measure their lengths to the nearest inch. The sample space for this random experiment may be $S_2 = \{1, 2, 3, \dots\}$. On the other hand, a more appropriate sample space could be $S_3 = \{x \mid x > 0\}$, which is read as "the set of all x such that $x > 0$." This sample space reflects the fact that length is not always, in fact is seldom, an integer. Whether S_2 or S_3 should be used in a particular problem depends on the nature of the measurement process. If decimals are to be used, we need S_3. If only integers are to be used, S_2 will suffice. The point is that sample spaces for a particular experiment are not unique and must be selected to provide all pertinent information for a given situation.

Consider the die rolling experiment again. Suppose we are interested in obtaining a 1 on a single roll; that is, rolling a 1 is an event of interest. Other possible events are rolling an even number, rolling a number less than 4, and so on. Formally, an *event* is any subset of the sample space. Because the null set (the set with no outcomes and often denoted by ϕ) is a subset of every set, the null set is an event associated with every sample space. Similarly, because every set is a subset of itself, the sample space is an event in every experiment. If the outcome of an experiment is contained in the event E, then the event E is said to have occurred. Events are usually denoted with capital letters at the beginning of the alphabet.

Returning to the die rolling experiment, let A be the event of rolling a 1; that is, $A = \{1\}$. Define $B = \{2, 4, 6\}$, the event of obtaining an even number on the roll. And define $C = \{1, 2, 3\}$, the event of rolling a

number less than 4. If the die is rolled, and a 2 appears on the upper face, then events B and C have both occurred. Note: It is enough that one of the outcomes in the event occurs for the event to occur; not all elements of the event must be observed (in fact, they cannot).

Example

A highway patrolman randomly selects a car on the interstate and measures its speed by radar. The patrolman may decide to let the car continue, issue a warning, or issue a ticket.

1. Give the sample space.
2. List all events.
3. The patrolman issued a ticket. Which event(s) has (have) occurred?
4. Are the outcomes for this random experiment equally likely?

Solution

1. $S = \{C, W, T\}$ where C represents the outcome of the car continuing, W is the outcome of the patrolman issuing a warning, and T is the outcome of the patrolman issuing a ticket.
2. Because there are three possible outcomes, there are 2^3 events: ϕ, $\{C\}$, $\{W\}$, $\{T\}$, $\{C,W\}$, $\{C,T\}$, $\{W,T\}$, S.
3. $\{T\}$, $\{C,T\}$, $\{W,T\}$, S.
4. It is very unlikely that all three outcomes are equally likely. For example, if a patrolman stops a car, it is generally more likely that a ticket and not a warning will be issued.

The probability of an event E, denoted by P(E), is the ratio of the number of outcomes favorable to E to the total number of outcomes in the sample space S.

► Practice

A movie theatre is showing four different movies at any one time. On a given night, one of the theatre's customers is selected at random, and the movie he attends is observed.

1. Give the sample space.
2. List all events.
3. The customer went to the third movie. Which event(s) has (have) occurred?
4. Are the outcomes for this random experiment equally likely?

► Definition of Probability

Now that sample space and events have been defined for a random experiment, we need to consider how to associate probabilities with all possible events. The *classical definition of probability* is as follows: The probability of an event E, denoted by P(E), is the ratio of the number of outcomes favorable to E to the total number of outcomes in the sample space S. For our die example, suppose we are interested in the probability of the event A, rolling an even number. The number of outcomes favorable to A is three because the outcomes "2," "4," and "6" are all possible outcomes that are even. The total number of outcomes in the sample space is six. Thus, by the classical definition of probability,

$$P(A) = \frac{number\ of\ outcomes\ favorable\ to\ A}{number\ of\ outcomes\ in\ the\ sample\ space}$$

$$= \frac{3}{6} = \frac{1}{2}.$$

It is important to note that this method for calculating probabilities is appropriate only when the outcomes of an experiment are *equally likely*. Further, both the event A and the sample space S must have a finite number of outcomes.

Sample spaces for many experiments can be constructed so that the outcomes are equally likely. However, this is not always possible. Suppose that an airline wants to know what proportion of the calls it receives on its reservation line results in a reservation. The sample space consists of only two outcomes: A reservation is made and a reservation is not made. It is very unlikely that these outcomes are equally likely. To determine what the probability is that a randomly selected call will result in a reservation, the company could record calls and determine whether or not each call resulted in a reservation. Then, based on a large number of calls, the proportion resulting in reservations would provide an estimate of the probability of interest. This is an example of the relative frequency definition of probability. More formally, for the *relative frequency definition of probability*, the probability of an event E, denoted by P(E), is defined to be the relative frequency of occurrence of E in a very large number of trials of a random experiment; that is, if the number of trials is quite large:

$$P(E) = \frac{number\ of\ times\ E\ occurs}{number\ of\ trials}$$

The probability of an event *E*, denoted by *P(E)*, is defined to be the relative frequency of occurrence of *E* in a very large number of trials of a random experiment.

It is important that there be a very large number of trials for the relative frequency to accurately reflect the probability of interest. The reason for this is that the relative frequency can fluctuate quite a bit for a small number of trials. For example, suppose that the probability that a randomly selected call to the airline results in a reservation is 0.4. Suppose we use the relative frequency definition of probability to find this probability. In Figure 9.1, we graph the relative frequency for a particular set of trials as the number of trials increases up to 100. Notice how much fluctuation occurs with only a few trials, but that the relative frequency tends to stabilize around 0.4 as the number of trials increases.

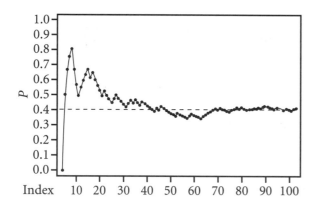

Figure 9.1

Looking both at the classical and relative frequency definitions of probability, we can observe some characteristics that are true for all probabilities.

- **Axiom 1:** Every probability must be between 0 and 1; that is, for any probability p, $0 \leq p \leq 1$.
- **Axiom 2:** Because the sample space is the set of all possible outcomes, one of the outcomes in *S*, and thus, the event *S*, must occur any time the experiment is conducted. Therefore, the probability of the sample space is one (i.e., $P(S) = 1$).
- **Axiom 3:** If two events *E* and *F* are disjointed (have no outcomes in common), then $P(E \text{ or } F) = P(E) + P(F)$ because, using the classical definition of probability,

$$P(E \text{ or } F) = \frac{\text{number of outcomes favorable to } E \text{ or } F}{\text{number of outcomes in the sample space}}$$

$$= \frac{\text{number of outcomes favorable to } E}{\text{number of outcomes in the sample space}}$$

$$+ \frac{\text{number of outcomes favorable to } F}{\text{number of outcomes in the sample space}}$$

$$= P(E) + P(F)$$

The third characteristic also holds if there are more than two disjointed sets.

These three characteristics are called the *axioms*, or rules, of probability. Using the three axioms of probability, more results of probability can be obtained. Three of these follow.

■ **Result 1:** The null event has no outcomes in the sample space and thus has probability zero of occurring (i.e., $P(\phi) = 0$).

■ **Result 2:** For any event E, the complement of E is the set of all outcomes in the sample space that are not in E. The probability of the complement of E is $1 - P(E)$. To see this, note that $P(E) + P(\text{not } E) = 1$, because every outcome in the sample space is in E or the complement of E. By subtraction, we then have

$$P(E) = 1 - P(\text{not } E).$$

■ **Result 3:** For any events E and F, $P(E \text{ or } F) = P(E) + P(F) - P(E \text{ and } F)$. To see this, notice that if we add the probability of E to the probability of F as we did for axiom 3, we have added the probability associated with the outcomes that are in both events twice, once for E and once for F. Thus, we must subtract out the probability of the outcomes the two events have in common to correctly compute the probability of E or F.

Using these basic ideas, we can find probabilities for a number of experiments.

Example

A candy store manager has kept a record of the purchases his customers make. He has determined the probability that a randomly selected customer will purchase gum is 0.4; the probability that a randomly selected customer will buy chocolate candy is 0.7; and the probability the customer purchases both is 0.2.

1. What is the probability that a randomly selected customer buys gum or chocolate candy?
2. What is the probability that a randomly selected customer will not buy gum?
3. What is the probability that a randomly selected customer will buy neither gum nor chocolate candy?

Solution

Let $G =$ the event a customer purchases gum and $C =$ the event a customer purchases chocolate candy. We are told $P(G) = 0.4$, $P(C) = 0.7$, and $P(G \text{ and } C) = 0.2$. Hint: It always helps to write down what is given using proper notation.

1. We want to find the probability of G or C occurring, so we will use axiom 6:

$$P(G \text{ or } C) = P(G) + P(C) - P(G \text{ and } C)$$
$$= 0.4 + 0.7 - 0.2 = 0.9$$

2. Not buying gum is the complement of buying gum. Thus:

$$P(\text{not } G) = 1 - P(G) = 1 - 0.4 = 0.6$$

3. The event of not buying either gum or chocolate candy is the complement of buying one or the other or both. That is:

$$P(\text{not } G \text{ and not } C) = 1 - P(G \text{ or } C)$$
$$= 1 - 0.9 = 0.1$$

In general, the *conditional probability* of the event E given that the event F has occurred is:

$$P(E|F) = \frac{P(E \text{ and } F)}{P(F)}$$

► Practice

Seniors at a high school have the opportunity to choose several elective courses, including photography and drawing. The probability that a randomly selected senior is taking photography is 0.50 and the probability that a randomly selected senior is taking drawing is 0.30. The probability that a randomly selected senior is taking both classes is 0.10.

5. What is the probability that a randomly selected senior is taking photography or drawing?

6. What is the probability that a randomly selected senior is not taking photography?

7. What is the probability that a randomly selected senior is taking neither photography nor drawing?

► Conditional Probability and Independence

Sometimes, knowing that one event has occurred changes our knowledge about the likelihood that another event has occurred. For example, suppose a family with two children is moving into the house across the street. What is the probability that the family has two girls? Using B to denote a boy and G to represent a girl, the sample space for this experiment is $S = \{BB, BG, GB, GG\}$. Note that both BG and GB are outcomes in the sample space because BG represents the outcome that the older child is a boy, GB implies the older child is a girl. For all of the outcomes in the

sample space to be equally likely, both must be listed. The probability of two girls is $\frac{1}{4}$ because only one of the four outcomes in the sample space has two girls.

Now suppose that we learn that the family has at least one girl. We know that the family does not have two boys, so the sample space is reduced to $S_2 = \{BG, GB, GG\}$. Each of the outcomes remains equally likely, so the probability of two girls knowing that the family has at least one girl is $\frac{1}{3}$.

If the number of possible outcomes is small, as was the case in this example, it is not difficult to determine both the sample space for the experiment and the reduced sample space given some information. However, as the size of the sample space increases, this approach quickly becomes impractical. We need a general approach to determining conditional probabilities. In general, for any event F such that $P(F) > 0$, the *conditional probability* of the event E given that the event F has occurred is:

$$P(E|F) = \frac{P(E \text{ and } F)}{P(F)}$$

For our family with two children, E would be the event of two girls and F would be the event of at least one girl. If a family must have two girls *and* at least one girl, then they have two girls (E cannot occur if the family has only one girl). Thus, the probability of E and F both occurring is $\frac{1}{4}$, the probability of two girls. The probability F occurs is $\frac{3}{4}$ because three of the four outcomes in the sample space have at least one girl.

Therefore, $P(E|F) = \dfrac{\frac{1}{4}}{\frac{3}{4}} = \dfrac{1}{3}$, which agrees with our earlier result based on the reduced sample space.

Sometimes, knowledge that one event has occurred does not change the probability that another event has occurred; that is, $P(E|F) = P(E)$. Now, let us rewrite $P(E|F) = \dfrac{P(E \text{ and } F)}{P(F)}$ as $P(E \text{ and } F) = P(E|F)P(F)$. If E and F are independent, $P(E|F) = P(E)$ and thus $P(E \text{ and } F) = P(E)P(F)$. When $P(E \text{ and } F) = P(E)P(F)$ or, equivalently, $P(E|F) = P(E)$, the events E and F are said to be *independent events* because knowledge of whether or not the event F has occurred has no effect on the probability that E has occurred. Further, given that E and F are independent events, $P(E \text{ and } F) = P(E)P(F)$ and $P(E|F) = P(E)$.

Example

In a housing development, the steering committee wanted to examine the need for pet and child play areas. The head of each household in the division was asked how many children lived in the household and whether or not they had a dog. The results are shown in Table 9.1.

Table 9.1 Number of children and dogs per family

DOG	NUMBER OF CHILDREN			TOTALS
	0	1	2 or more	
Yes	12	46	74	132
No	18	48	58	124
Totals	30	94	132	256

1. What is the probability that a randomly selected household in this subdivision has a dog?
2. What is the probability that a randomly selected household in this subdivision has children?
3. What is the probability that a randomly selected household in this subdivision has a dog and one child?
4. What is the probability that a randomly selected household in this subdivision has a dog given that one child is in it?
5. Is the event of having a dog independent of having one child for households in this neighborhood?

Solution

Let D = event the household has a dog. Let A = event the household has no children, B = event the household has one child, C = event the household has at least two children, and E = event the household as at least one child.

1. Since 132 of the 156 households have dogs, $P(D) = \dfrac{132}{256}$.

2. A household has children if it has one or more children. Thus,
$$P(E) = P(B) + P(C) = \dfrac{94}{256} + \dfrac{132}{256} = \dfrac{226}{256}$$
is the probability that the randomly selected household has at least one child.

3. Forty-six households had both a dog and one child, so $P(D \text{ and } B) = \dfrac{46}{256}$.

4. Of the 94 households with one child, 46 had a dog. Therefore, $P(D|B) = \dfrac{46}{94}$.

5. Having a dog is not independent of having one child in this housing division because
$$P(D) = \dfrac{132}{256} \neq \dfrac{46}{94} = P(D|B).$$

▶ Practice

A high school athletic director wanted to know whether participation in high school sports differed with grade. Using school records, he determined how many students in each grade did not play any sport and how many played at least one sport. The results are shown in Table 9.2.

Table 9.2 HS students playing sports by grade level

SPORT	FRESHMAN	SOPHOMORE	JUNIOR	SENIOR	TOTALS
Yes	192	131	92	80	495
No	90	122	122	127	461
Totals	282	253	214	207	956

Let P = the event that the student plays at least one high school sport, A = the event that the student is a freshman, B = the event that the student is a sophomore, C = the event that the student is a junior, and D = the event that the student is a senior.

8. What is the probability that a randomly selected student at this high school plays at least one high school sport?

9. What is the probability that a randomly selected student is an upperclassman (junior or senior)?

10. What is the probability that a randomly selected student is a junior who plays at least one high school sport?

11. What is the probability that a randomly selected student plays at least one high school sport given that he or she is a junior?

12. Is the probability of a student playing at least one high school sport independent of he or she being a junior?

▶ Theorem of Total Probability

Suppose that the sample space is partitioned into n mutually exclusive and exhaustive events, E_i, i = 1, 2, . . . , n, as in Figure 9.2. By *mutually exclusive*, we mean that the events have no outcomes in common. By *exhaustive*, we mean that, when combining the outcomes from all events, we have all points in the sample space. Now suppose that we are interested in determining the probability of the event A. It may be difficult to find the probability of A directly, but we can sometimes use the partitioning of the sample space to make this easier. The *Theorem of Total Probability* states that, if E_i, i = 1, 2, . . . , n are disjoint events, with $P(E_1) + P(E_2) + \ldots + P(E_n) = 1$, then for any event F:

$$P(F) = P(F \text{ and } E_1) + P(F \text{ and } E_2) + \ldots + P(F \text{ and } E_n) = P(F|E_1)P(E_1) + P(F|E_2)P(E_2) + \ldots + P(F|E_n)P(E_n)$$

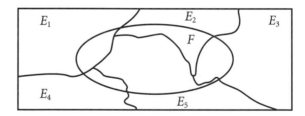

Figure 9.2

Theorem of Total Probability

If E_i, $i = 1, 2, \ldots, n$ are disjoint events, with $P(E_1) + P(E_2) + \ldots + P(E_n) = 1$, then for any event F:

$P(F) = P(F \text{ and } E_1) + P(F \text{ and } E_2) + \ldots + P(F \text{ and } E_n) = P(F|E_1)P(E_1) + P(F|E_2)P(E_2) + \ldots + P(F|E_n)P(E_n)$

Example

Suppose that a company uses a specialized part in its manufacturing process, and the part can be purchased from only three suppliers. The company consistently purchases 60%, 30%, and 10% of its stock of this part from suppliers A, B, and C, respectively. A portion of the parts from each supplier is defective. The company has determined that 1%, 3%, and 8% of the parts purchased from suppliers A, B, and C, respectively, are defective. What is the probability that a randomly selected part from the company's stock is defective?

Solution

When working probability problems, it is best to start by presenting what is known using symbols. Notice that we know the probability that a randomly selected part comes from each of the suppliers. Let E_i be the event that a randomly selected part from the company's stock is from supplier i, $i = A, B,$ or C. Then, for a randomly selected part, $P(E_A) = 0.6$, $P(E_B) = 0.3$, and $P(E_C) = 0.1$. Notice that E_A, E_B, and E_C are disjoint (a randomly selected part cannot come from two different suppliers) and exhaustive (any randomly selected part must have come from one of these three suppliers). Consequently, $P(E_A \text{ and } E_B) = P(E_A \text{ and } E_C) = P(E_B \text{ and } E_C) = 0$ (because any two events are disjoint, the probability of both occurring is zero), and $P(E_A) + P(E_B) + P(E_C) = 1$.

Let D be the event that a randomly selected part is defective. Notice that D overlaps with E_A, E_B, and E_C because it is possible for a defective part to come from any supplier. Further, although we do not know the probability of D, we do know the probability of D if we are given the supplier. That is, we know $P(D|E_A) = 0.01$, $P(D|E_B) = 0.03$, and $P(D|E_C) = 0.08$. Therefore, if we add together the probabilities that a part was defective and from supplier A, defective and from supplier B, and defective and from supplier C, we would have the probability of a defective part. We can represent this in symbols as

$$P(D) = P(D \text{ and } E_A) + P(D \text{ and } E_B) + P(D \text{ and } E_C)$$
$$= P(D|E_A)P(E_A) + P(D|E_B)P(E_B) + P(D|E_C)P(E_C)$$
$$= 0.01(0.6) + 0.03(0.3) + 0.08(0.1)$$
$$= 0.023$$

If E_i, $i = 1, 2, \ldots, n$ are disjoint events with $P(E_1) + P(E_2) + \ldots + P(E_n) = 1$, then for any event F,

$$P(E_i|F) = \frac{P(F|E_i)P(E_i)}{P(F|E_1)P(E_1) + P(F|E_2)P(E_2) + \ldots + P(F|E_n)P(E_n)}.$$

Thus, the probability that a randomly selected part is defective is 0.023.

▶ Practice

13. The staff at a large high school is 35% male and 65% female. When asked if they liked the cafeteria food at the high school, 67% of the men on staff said they did, and 59% of the women on staff said they liked the food. What is the probability that a randomly selected staff member at the high school likes the school's cafeteria food?

▶ Bayes' Theorem

As with the Theorem of Total Probability, suppose that the sample space is partitioned into n mutually exclusive and exhaustive events, E_i, $i = 1, 2, \ldots, n$. Also, assume that for an event F on the same sample space S, $P(F|E_i)$ is known for all i. However, instead of finding $P(F)$, we want to find $P(E_i|F)$. Bayes' Theorem says that, if E_i, $i = 1, 2, \ldots, n$ are disjoint events with $P(E_1) + P(E_2) + \ldots + P(E_n) = 1$, then for any event F,

$$P(E_i|F) = \frac{P(F|E_i)P(E_i)}{P(F|E_1)P(E_1) + P(F|E_2)P(E_2) + \ldots + P(F|E_n)P(E_n)}.$$

Example

Suppose that, for the same manufacturer and parts suppliers in the last example, the manufacturer selects a part at random and determines it to be defective. What is the probability that it came from supplier A? Notice that we want to find $P(E_A|D)$. If we begin by using the definition of conditional probability, we have

$$P(E_A|D) = \frac{P(E_A \text{ and } D)}{P(D)}.$$

Using the properties of conditional probability and the Theorem of Total Probability, we can rewrite the conditional probability of E_A given D as

$$P(E_A|D)$$
$$= \frac{P(D|E_A)P(E_A)}{P(D|E_A)P(E_A) + P(D|E_B)P(E_B) + P(D|E_C)P(E_C)}$$
$$= \frac{0.01(0.6)}{0.01(0.6) + 0.03(0.3) + 0.08(0.1)}$$
$$= \frac{6}{23}$$

Of course, because we had already found the probability that a randomly selected part is defective, it was not necessary to expand the denominator using the Theorem of Total Probability. Finding the probability that the supplier was E_A given that the part was defective is an example of the application of Bayes' Theorem.

▶ Practice

14. Suppose that, for the same high school mentioned in the previous practice problem, we select a staff member at random and determine that this person likes the cafeteria food. What is the probability that the staff member is a man?

▶ In Short

Probability is the foundation of statistics. In this lesson, we learned that the sample space is the set of all possible outcomes of an experiment, and an event is any subset of the sample space. The classical definition and the relative frequency definition of probability were compared. Conditional probabilities and independent events were discussed. Finally, the Theorem of Total Probability and Bayes' Theorem were presented as useful methods of finding probabilities of interest.

Discrete Probability Distributions

LESSON SUMMARY

In the previous chapter, we discussed how to attach probabilities to events and we discussed a few properties of probabilities. Because some probability distributions occur frequently in practice, they have been given specific names. In this lesson, we will discuss three discrete probability distributions: the Bernoulli, the binomial, and the geometric distributions.

▶ Bernoulli Distribution

Suppose we flip a fair coin and observe the upper face. The sample space may be represented as $S = \{$Head, Tail$\}$. Suppose we define $X = 1$ if a head is on the upper face and 0 if a tail is on the upper face. X is an example of a random variable. We have used the term *random variable* somewhat loosely in earlier lessons. Formally, a random variable X assigns a numerical result to each possible outcome of a random experiment. If X can assume a finite or countably infinite number of values, then X is a discrete random variable; otherwise, X is a continuous random variable. In this chapter, we will consider the distributions of some discrete random variables.

The probability function $P(X = x)$, or $p(x)$, assigns a probability to each possible value of X. Because these are probabilities, $0 \leq p(x) \leq 1$ for all $X = x$. Further, if we sum over all possible values of X, we must get one (i.e., $\sum_x p(x) = 1$).

Bernoulli Trial

A Bernoulli trial is any random experiment that has only two possible outcomes.

A discrete probability function is any function that satisfies the following two conditions: (1) The probabilities are between 0 and 1 and (2) the probabilities sum to one. As an illustration, let $X = -1, 0,$ or 1 if the stock market goes down, up, or stays the same, respectively, on a given day. The probabilities associated with the particular outcomes of X change from day to day. However, suppose for a given day, they are as follows:

Table 10.1 Probability of stock market change

X	−1	0	1
P(X)	0.3	0.1	0.6

Each probability is between 0 and 1, and the sum of the probabilities is one. Thus, this is a valid probability function. The graph of the distribution is given in Figure 10.1.

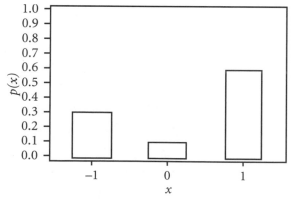

Figure 10.1

For the moment, we are going to focus on studies in which each observation may result in one of two possible outcomes. Flipping the coin is one such study as each flip will result in either a head or a tail. In an orchard, each piece of fruit either has or has not been damaged by insects. The television set tested at the factory either works or it does not. A person has a job or does not have a job. In each case, there are only two possible outcomes; one outcome may be labeled a success and the other a failure. A *Bernoulli trial* is any random experiment that has only two possible outcomes. For a Bernoulli trial, let X be a random variable defined as follows:

$$X = \begin{cases} 1, \textit{if a success is observed} \\ 0, \textit{if a failure is observed} \end{cases}$$

The choice of which outcome is considered a success and which is considered a failure is arbitrary. It is only important to clearly state for which outcome $X = 1$ and for which $X = 0$. The probability of success is denoted by p where $0 < p < 1$. Because there are only two outcomes, the probability of a success and the probability of a failure must sum to 1. Thus, the probability of a failure is $1 - p$. We can present the probability distribution of the Bernoulli random variable as shown in Table 10.2.

Table 10.2 Probability distribution of the Bernoulli random variable

x	0	1
p(x)	1 − p	p

If x is the number of successes in n independent Bernoulli trials, each with the probability of p success, x is a binomial random variable.

▶ Binomial Distribution

Seldom are we satisfied with performing one Bernoulli trial. Instead, we want to conduct multiple Bernoulli trials, observing the outcome of each one. Suppose we have n *independent* Bernoulli trials, each with the probability p of success. Let X be the number of successes observed in the n trials. Then X is a binomial random variable. The probability of x successes in n trials may be written as:

$$p(x) = \binom{n}{x} p^x (1 - p)^{n-x}, x = 0, 1, 2, \ldots, n$$

In the above equation, $\binom{n}{x} = C_r^n = \dfrac{n!}{x!(n - x)!}$ is the number of ways to choose x items from n and is called the number of combinations of n things taken x at a time. (Recall that $n! = n(n - 1)(n - 2) \ldots (1)$ so that $5! = 5(4)(3)(2)(1) = 120$.) We will consider some examples of the binomial distribution and then provide an explanation of why the probabilities are computed as stated here.

Suppose we randomly select 25 apples from the orchard and count how many are damaged. If one apple being damaged has no effect on whether or not the next selected apple is damaged, then the number of damaged apples would have a binomial distribution with $n = 25$ and p equal to the proportion of damaged apples in the orchard.

In a quality control study, we could randomly choose 50 television sets from the production line and carefully test each to determine how many are defective. If we assume that whether or not one television is defective is independent of the next television being defective, then the number of defective televisions would have a binomial distribution with $n = 50$ and p equal to the proportion of defective sets produced during the period of the study.

To understand why the probabilities associated with the binomial random variable are as given in this way, first suppose we flip a coin twice and observe the number of heads. Each flip may be considered to be a Bernoulli trial. Because the outcome on one flip does not affect the outcome on the next flip, the two trials are independent. Let X be the random variable denoting the number of heads observed on the two flips. Then $X = 0, 1,$ or 2. In Figure 10.2, we have a tree diagram representing this experiment. Notice that each flip results in a set of two branches, one representing heads and the other tails. The tree has four terminal branches, representing the outcomes $HH, HT, TH,$ and TT, where H represents heads and T represents tails. The random variable X assigns 2 to HH, 1 to HT, 1 to TH, and 0 to TT. Because all four outcomes are equally likely, we have the following probability distribution for X.

Table 10.3 Probability of heads and tails

x	0	1	2
$p(x)$	$\dfrac{1}{4}$	$\dfrac{1}{2}$	$\dfrac{1}{4}$

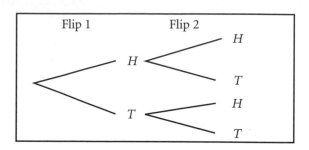

Figure 10.2

Notice that because the two flips are independent, the probability of two heads is $p(2) = p^2$. Here, $p = \frac{1}{2}$, so $p^2 = \frac{1}{4}$. Other outcomes can be computed in the same manner. Also note that the values of X are not equally likely. Because there are two ways to obtain one head (*HT* and *TH*) and only one way to obtain either zero heads (*TT*) or two heads (*HH*), the probability that $X = 1$ is twice that of $X = 0$ or $X = 2$. Thus, we have $2p(1 - p)$ as the probability of one head.

When experiments increase in size, as would be the case if we flipped the coin 40 times, it becomes unreasonable to construct a tree diagram or to list all possible outcomes. We need a general way of counting the number of ways to get x successes in n trials. The number of ways to choose x items from n is $\frac{n!}{x!(n - x)!}$ and is called the number of combinations of n things taken r at a time.

For the coin flipping example, we have $n = 2$. For $X = 0$, we have $\frac{2!}{0!2!} = \frac{2}{1(2)} = 1$. When $X = 1$, $\frac{2!}{1!1!} = \frac{2}{1(1)} = 2$. When $X = 2$, this function is again 1. Thus, we have accurately counted the number of ways to get 0, 1, or 2 heads. The probability of any particular sequence of heads and tails is $p^x(1 - p)^{n-x}$ because we have x successes, each with probability p, and $(n - x)$ failures, each with probability $1 - p$. Thus, the probability of X successes in n trials may be written as

$$p(x) = \binom{n}{x}p^x(1 - p)^{n-x}, x = 0, 1, 2, \ldots, n, \quad \text{which}$$
is the probability function given earlier.

Example

A student knows that the test tomorrow will have ten true-false questions on it. She decides to flip a coin and to mark true if the upper face is a head and mark false if the upper face is a tail. She will repeat the process for each question. What is the probability that she will miss every question?

Solution

Because the student flips the coin to determine the response, she has a probability of $p = \frac{1}{2}$ of getting each question correct. The flips are independent, so whether she gets a question correct is independent of whether she gets any other question correct. The probability that the student does not get any question correct is then:

$$p(0) = \binom{10}{0}\left(\frac{1}{2}\right)^0\left(1 - \frac{1}{2}\right)^{10-0}$$
$$= \left(\frac{1}{2}\right)^{10} = \frac{1}{1,024} = 0.000977$$

It is very likely that she will get at least some of the questions correct.

Example

In a study, dogs were trained to detect the presence of bladder cancer by smelling urine (see *USA Today*, September 24, 2004). During training, each dog was presented with urine specimens from healthy people, those from people with bladder cancer, and those from people sick with unrelated diseases. The dog was to lie down by any urine specimen from a person with bladder cancer. After training, each dog was presented with seven urine specimens, only one of which came from a person with cancer. The specimen that the dog laid down beside was recorded. If the dog identified the urine specimen from a person with cancer, the test was considered a success; otherwise, it was a failure. Each dog repeated the test nine times. If a dog cannot detect the presence of bladder cancer by smelling the urine, what is the probability he will identify the specimen with cancer in at least eight of the trials?

Solution

At first glance, this may not seem like a series of Bernoulli trials. However, notice that for each trial, there are two possible outcomes: choosing the specimen associated with cancer or choosing a specimen of an individual without cancer. Because there are six noncancer specimens and only one with cancer, the probability of success is $p = \dfrac{1}{7}$ if the dog cannot detect bladder cancer in urine and so chooses one at random. The trials are independent. Thus, the probability of the dog making the correct identification in at least eight of the nine trials is:

$$
\begin{aligned}
p(8) &+ p(9) \\
&= \binom{9}{8}\left(\frac{1}{7}\right)^8\left(1 - \frac{1}{7}\right)^1 + \binom{9}{9}\left(\frac{1}{7}\right)^9\left(1 - \frac{1}{7}\right)^0 \\
&= 9\left(\frac{6}{7^9}\right) + \frac{1}{7^9} \\
&= 0.00000136
\end{aligned}
$$

It would be very unlikely for a dog to detect the urine sample from the person with bladder cancer in at least eight of nine trials by chance alone.

Practice

1. Sixteen playing cards were put into a hat; eight red and eight black. A person is asked to blindly take out eight cards from the hat. Each time the person takes out a card, the value of the card is recorded as red or black and then the card is placed back into the hat. What is the probability that all the cards the person draws are black?

2. If we roll a dice 15 times, what is the probability that we will roll a 6 at least 13 times?

Geometric Distribution

For the binomial distribution, we had n independent Bernoulli trials, each with the probability p of success. Suppose now that we have independent Bernoulli trials each with the probability p of success. However, let X be the number of failures prior to the first success. X is a geometric random variable, and the probability that $X = x$ is $p(x) = (1 - p)^x p, x = 0, 1, 2, \ldots$

To understand why the probabilities are computed in this manner, first note that, if p is the probability of success, $(1 - p)$ is the probability of failure. Further, there is only one way to have x failures prior to the first success; otherwise, we will get a success before the xth failure.

A geometric random variable is sometimes defined as the number of trials needed to obtain the first success instead of as the number of failures prior to the first success. Both definitions are valid, but the probability function differs slightly for the two, so it is important to read the definition carefully when moving from one source to the next.

Example

If we repeatedly flip a fair coin, what is the probability that we will get the first head on the third flip?

Solution

Because we have a fair coin, the probability of a head on each flip is $p = \frac{1}{2}$. If the first head is on the third flip, we had two tails (or failures) prior to this first success. Thus, the probability that the number of failures $X = 2$ prior to the first success is

$$p(2) = \left(1 - \frac{1}{2}\right)^2 \left(\frac{1}{2}\right) = \left(\frac{1}{2}\right)^3 = \frac{1}{8} = 0.125.$$

▶ Practice

3. Referring back to the red and black card example, what is the probability that the first black card will be drawn on the fourth trial?

▶ In Short

Three common discrete distributions are the Bernoulli, the binomial, and the geometric. The Bernoulli arises when a trial in an experiment has two possible outcomes, commonly referred to as a success and a failure. If we conduct n independent Bernoulli trials, each with the probability p of success, the number of successes is a binomial random variable. If we conduct independent Bernoulli trials, each with the probability p of success, the number of failures prior to the first success is a geometric random variable.

LESSON 11 ▶ Continuous Probability Distributions

LESSON SUMMARY

As was the case with discrete distributions, some continuous random variables are of particular interest. In this lesson, we will discuss two of these: the uniform distribution and the normal distribution. The normal distribution is particularly important because many of the methods used in statistics are based on this distribution. The reasons for this will become clearer as we work through the rest of the lessons.

▶ Uniform Distribution

In the first lesson, we learned that a continuous random variable has a set of possible values that is an interval on the number line. It is not possible to assign a probability to each point in the interval and still satisfy the conditions of probability set forth in Lesson 10 for discrete random variables. Instead, the probability distribution of a continuous random variable X is specified by a mathematical function $f(x)$ called the *probability density function* or just *density function*. The graph of a density function is a smooth curve. A *probability density function* (pdf) must satisfy two conditions: (1) $f(x) \geq 0$ for all real values of x and (2) the total area under the density curve is equal to 1. The graphs of three density functions are shown in Figure 11.1.

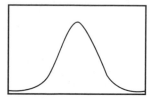
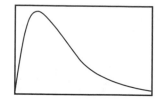
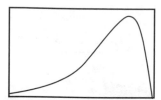

Figure 11.1

The probability that X lies in any particular interval is shown by the area under the density curve and above the interval. The following three events are frequently encountered: (1) $X < a$, the event that the random variable X assumes a value less than a; (2) $a < X < b$, the event that the random variable X assumes a value between a and b; and (3) $X > b$, the event that the random variable X is greater than b. We say that we are interested in the lower tail probability for (1) and the upper tail probability when using (3). The areas associated with each of these are shown in Figure 11.2.

Notice that the probability that $a < X < b$ may be computed using tail probabilities:

$$P(a < X < b) = P(X < b) - P(X < a).$$

If the random variable X is equally likely to assume any value in an interval (a, b), then X is a uniform random variable. The pdf is flat and is above the x-axis between a and b, and it is 0 outside of the interval. The height of the curve must be such that the area

under the density and above the x-axis is 1. Because this region is a rectangle, the area is the height times the width of the interval, which is $b - a$. Thus, the height must be $\dfrac{1}{(b - a)}$; that is, the pdf of a uniform random variable has the form

$$f(x) = \frac{1}{b - a}, \ a < x < b$$

$$= 0, \ otherwise.$$

A graph of the pdf is shown in Figure 11.3.

Example

A group of volcanologists (people who study volcanoes) has been monitoring a volcano's seismicity, or the frequency and distribution of underlying earthquakes. Based on these readings, they believe that the volcano will erupt within the next 24 hours, but the eruption is equally likely to occur any time within that period. What is the probability that it will erupt within the next eight hours?

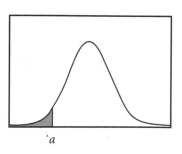
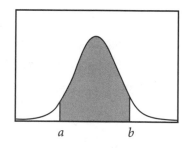
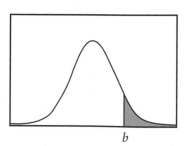

Figure 11.2

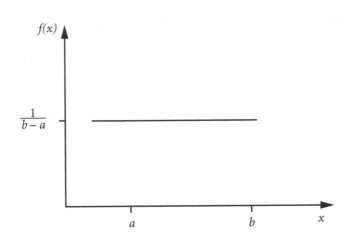

Figure 11.3

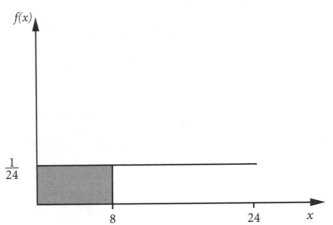

Figure 11.4

Solution

Define X = the time until the eruption of the volcano. X has positive probability over the interval $(0,24)$ because the volcano will erupt during that time interval. Because the length of the interval is $24 - 0 = 24$, the height of the density curve must be $\frac{1}{24}$ for the area under the density and above the x-axis to be one. That is, the pdf is

$$f(x) = \frac{1}{24}, \quad 0 < x < 24$$

$$= 0, \quad otherwise.$$

The probability that the volcano will erupt within the next eight hours is equal to the area under the curve and above the interval $(0,8)$ as shown in Figure 11.4. This area is $8\left(\frac{1}{24}\right) = \frac{1}{3}$.

In the previous example, notice that the area is the same whether we have $P(0 < X < 8)$ or $P(0 \le X < 8)$ or $P(0 < X \le 8)$ or $P(0 \le X \le 8)$. Unlike discrete random variables, whether the inequality is strict or not, the probability is the same for the continuous random variables. This also correctly implies that, for con-

tinuous random variables, the probability that the random variable equals a specific value is 0.

▶ Practice

1. Lindsay has a date tonight with her boyfriend Andrew. Andrew told Lindsay that he would pick her up at 7 P.M. to go to dinner. Andrew always arrives to pick Lindsay up within 20 minutes after (never before) the time he said he would, and he is equally likely to arrive anytime during those 20 minutes. What is the probability Andrew will show up no more than 10 minutes late?

Normal probability distributions are continuous probability distributions that are bell shaped and symmetric. They are also known as *Gaussian distributions* or *bell-shaped curves*.

▶ Normal Distribution

The normal distribution is perhaps the most widely used probability distribution, largely because it provides a reasonable approximation to the distribution of many random variables. It also plays a central role in many of the statistical methods that will be discussed in later lessons. *Normal probability distributions* are continuous probability distributions that are bell shaped and symmetric as displayed in Figure 11.5. The distribution is also called the *Gaussian distribution* or the *bell-shaped curve*.

The normal distribution has two parameters: the mean μ and the standard deviation σ. The notation $X \sim N(\mu,\sigma)$ means that "X is normally distributed with a mean of μ and a standard deviation of σ." The distribution is symmetric about the mean. The mean, median, and mode are all equal. The mean is often referred to as the location parameter because it determines where the distribution is centered. The standard deviation determines the spread of the distribution. The effect of the mean and standard deviation on the normal distribution is displayed in Figure 11.6.

For any normal distribution, about 68% of the observations are within one standard deviation of the mean. About 95% and 99.7% of the observations are, respectively, within two and three standard deviations of the mean.

It is important to remember that, although the location and spread may change, the area under the curve and above the x-axis is always 1. Unfortunately, the probabilities associated with intervals cannot be

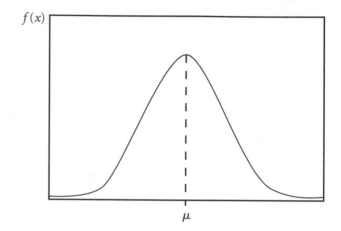

Figure 11.5

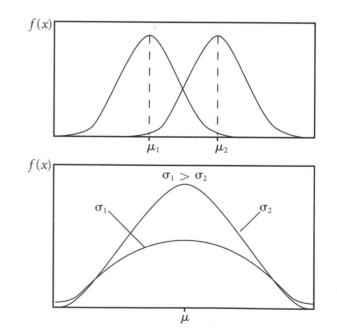

Figure 11.6

computed easily as with the uniform distribution. To overcome this difficulty, we rely on a table of areas for a reference of normal distribution called the *standard normal distribution*. The standard normal distribution is the normal distribution with $\mu = 0$ and $\sigma = 1$. It is customary to use the letter z to represent a standard normal random variable.

We will first learn to compute probabilities for a standard normal random variable and then learn how to find them for any random variable. We will also want to be able to determine extreme values of z, such as the value that only 5% of the population exceeds or the value that 1% of the population is less than. To find either probabilities or extreme values, we need a table of standard normal curve areas, or we need a calculator or computer that can be used to find these values. Here, we will restrict ourselves to the use of tables. The standard normal table used here in Table 11.1 tabulates the probability of observing a value less than or equal to z (see Figure 11.7).

Graphs are extremely useful tools to help us understand what values we are searching for. We will do this for each problem we work.

Example
Find $P(z < 1.32)$.

Solution
Using the standard normal table, we find the row with 1.3 in the z column and move along that row to the 0.02 column to find 0.9066. Thus, $P(z < 1.32) = 0.9066$. Figure 11.8 shows the graphic image of this.

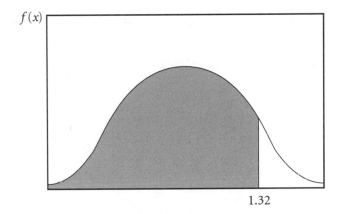

Figure 11.8

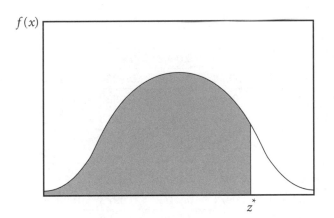

Figure 11.7

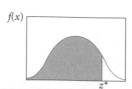

Table 11.1 Standard normal probabilities, $P(z < z^*)$

SECOND DIGIT OF z

z	0.00	0.01	0.02	0.03	0.04	0.05	0.06	0.07	0.08	0.09
0.0	0.5000	0.5040	0.5080	0.5120	0.5160	0.5199	0.5239	0.5279	0.5319	0.5359
0.1	0.5398	0.5438	0.5478	0.5517	0.5557	0.5596	0.5636	0.5675	0.5714	0.5753
0.2	0.5793	0.5832	0.5871	0.5910	0.5948	0.5987	0.6026	0.6064	0.6103	0.6141
0.3	0.6179	0.6217	0.6255	0.6293	0.6331	0.6368	0.6406	0.6443	0.6480	0.6517
0.4	0.6554	0.6591	0.6628	0.6664	0.6700	0.6736	0.6772	0.6808	0.6844	0.6879
0.5	0.6915	0.6950	0.6985	0.7019	0.7054	0.7088	0.7123	0.7157	0.7190	0.7224
0.6	0.7257	0.7291	0.7324	0.7357	0.7389	0.7422	0.7454	0.7486	0.7517	0.7549
0.7	0.7580	0.7611	0.7642	0.7673	0.7704	0.7734	0.7764	0.7794	0.7823	0.7852
0.8	0.7881	0.7910	0.7939	0.7967	0.7995	0.8023	0.8051	0.8078	0.8106	0.8133
0.9	0.8159	0.8186	0.8212	0.8238	0.8264	0.8289	0.8315	0.8340	0.8365	0.8389
1.0	0.8413	0.8438	0.8461	0.8485	0.8508	0.8531	0.8554	0.8577	0.8599	0.8621
1.1	0.8643	0.8665	0.8686	0.8708	0.8729	0.8749	0.8770	0.8790	0.8810	0.8830
1.2	0.8849	0.8869	0.8888	0.8907	0.8925	0.8944	0.8962	0.8980	0.8997	0.9015
1.3	0.9032	0.9049	0.9066	0.9082	0.9099	0.9115	0.9131	0.9147	0.9162	0.9177
1.4	0.9192	0.9207	0.9222	0.9236	0.9251	0.9265	0.9279	0.9292	0.9306	0.9319
1.5	0.9332	0.9345	0.9357	0.9370	0.9382	0.9394	0.9406	0.9418	0.9429	0.9441
1.6	0.9452	0.9463	0.9474	0.9484	0.9495	0.9505	0.9515	0.9525	0.9535	0.9545
1.7	0.9554	0.9564	0.9573	0.9582	0.9591	0.9599	0.9608	0.9616	0.9625	0.9633
1.8	0.9641	0.9649	0.9656	0.9664	0.9671	0.9678	0.9686	0.9693	0.9699	0.9706
1.9	0.9713	0.9719	0.9726	0.9732	0.9738	0.9744	0.9750	0.9756	0.9761	0.9767
2.0	0.9772	0.9778	0.9783	0.9788	0.9793	0.9798	0.9803	0.9808	0.9812	0.9817
2.1	0.9821	0.9826	0.9830	0.9834	0.9838	0.9842	0.9846	0.9850	0.9854	0.9857
2.2	0.9861	0.9864	0.9868	0.9871	0.9875	0.9878	0.9881	0.9884	0.9887	0.9890
2.3	0.9893	0.9896	0.9898	0.9901	0.9904	0.9906	0.9909	0.9911	0.9913	0.9916
2.4	0.9918	0.9920	0.9922	0.9925	0.9927	0.9929	0.9931	0.9932	0.9934	0.9936
2.5	0.9938	0.9940	0.9941	0.9943	0.9945	0.9946	0.9948	0.9949	0.9951	0.9952
2.6	0.9953	0.9955	0.9956	0.9957	0.9959	0.9960	0.9961	0.9962	0.9963	0.9964
2.7	0.9965	0.9966	0.9967	0.9968	0.9969	0.9970	0.9971	0.9972	0.9973	0.9974
2.8	0.9974	0.9975	0.9976	0.9977	0.9977	0.9978	0.9979	0.9979	0.9980	0.9981
2.9	0.9981	0.9982	0.9982	0.9983	0.9984	0.9984	0.9985	0.9985	0.9986	0.9986
3.0	0.9987	0.9987	0.9987	0.9988	0.9988	0.9989	0.9989	0.9989	0.9990	0.9990
3.1	0.9990	0.9991	0.9991	0.9991	0.9992	0.9992	0.9992	0.9992	0.9993	0.9993
3.2	0.9993	0.9993	0.9994	0.9994	0.9994	0.9994	0.9994	0.9995	0.9995	0.9995
3.3	0.9995	0.9995	0.9995	0.9996	0.9996	0.9996	0.9996	0.9996	0.9996	0.9997
3.4	0.9997	0.9997	0.9997	0.9997	0.9997	0.9997	0.9997	0.9997	0.9997	0.9998
3.5	0.9998	0.9998	0.9998	0.9998	0.9998	0.9998	0.9998	0.9998	0.9998	0.9998

Example

Find $P(z > 1.32)$.

Solution

From the table, we find $P(z < 1.32)$ as we did in the previous example. Using some of the ideas of probability we learned earlier, we have $P(z > 0.32) = 1 - P(z \leq 1.32) = 1 - 0.9066 = 0.0934$. See Figure 11.9.

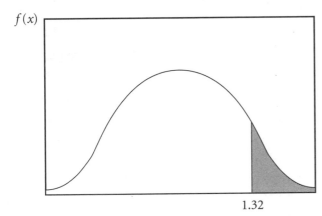

Figure 11.9

Example

Find $P(z < -0.5)$.

Solution

There are no negative z-values in the table, so we cannot look this up directly. Instead, we use the symmetry of the normal distribution to find the probability (see Figure 11.10). That is,

$$P(z < -0.5) = P(z > 0.5)$$
$$= 1 - P(z < 0.5)$$
$$= 1 - 0.6915$$
$$= 0.3085$$

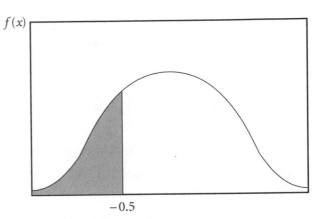

Figure 11.10

Example

Find $P(-1.45 < z < 0.76)$.

Solution

Figure 11.11 shows the solution.

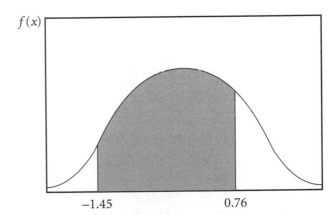

Figure 11.11

First, we notice that $P(-1.45 < z < 0.76) = P(z < 0.76) - P(z < -1.45)$. Now $P(z < 0.76)$ can be found directly from the table to be 0.7764. Using the symmetry of the normal distribution again, $P(z < -1.45) = P(z > 1.45) = 1 - P(z \leq 1.45) = 1 - 0.9265 = 0.0735$. Finally, $P(-1.45 < z < 0.76) = P(z < 0.76) - P(z < -1.45) = 0.7764 - 0.0735 = 0.7029$.

Example

Find the value z^* such that $P(z < z^*) = 0.75$.

Solution

This is different from the other problems we have considered. Instead of finding a probability, we are looking for a z-value. However, the same table will allow us to solve the problem. The difference is that we will look in the table for a probability and then find the z-value associated with the probability. Looking in the body of the table, we find the values 0.7486 and 0.7517, which are the closest to the 0.75 of interest. By looking at the corresponding row and column headings, we find that $P(z < 0.67) = 0.7486$ and $P(z < -0.68) = 0.7517$. Because 0.7486 is closer to 0.75 than 0.7517, we take $z^* = 0.67$. (Note: We could interpolate to find a more precise value of z^*, but we will not go through this process here.) See Figure 11.12.

Example

Find the value z^* such that $P(z > z^*) = 0.05$.

Solution

We need to have the probabilities in the form $P(z < z^*)$ to use the table. However, $P(z > z^*) = 1 - P(z \leq z^*)$. We can rewrite this as $P(z \leq z^*) = 1 - P(z > z^*) = 1 - 0.05 = 0.95$. That is, if 5% of the population values are greater than z^*, then 95% of the population values must be less than or equal to z^*. Thus, we look for 0.95 in the body of the table and find 0.9495 and 0.9505 corresponding to $z = 1.64$ and $z = 1.65$, respectively. Because 0.95 is exactly halfway between 0.9495 and 0.9505, we have $z^* = 1.645$. (This is the only time we don't just round to the closest value.) See Figure 11.13.

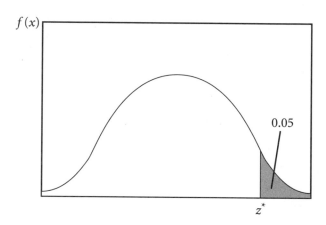

Figure 11.13

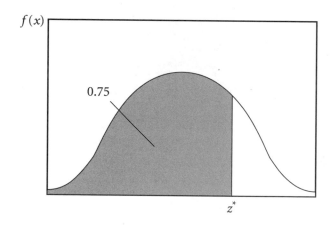

Figure 11.12

Example

Find the value z^* such that $P(z < z^*) = 0.01$.

Solution

Because the standard normal is symmetric about its mean 0, we know $P(z < 0) = 0.5$, we know that z^* must be less than 0. Also, because we have only positive values of z in the table, we cannot look for 0.01 directly in the table. However, again because of symmetry, we know that, if $P(z < z^*) = 0.01$, then $P(z > -z^*) = 0.01$. To use the table, we must find $P(z \leq -z^*) = 1 - P(z > -z^*) = 1 - 0.01 = 0.99$. Looking in the body of the table, we find 0.9898 and 0.9901, corresponding to $z = 2.32$ and $z = 2.33$, respectively, to be the closest to 0.99. Because 0.9901 is the closer of the two to 2.33, we find $z^* = -2.33$. See Figure 11.14.

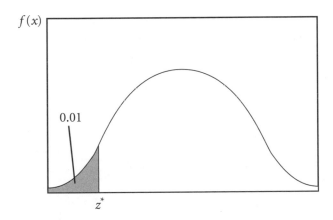

Figure 11.14

▶ Practice

2. Find $P(z < 0.75)$.
3. Find $P(z > 0.75)$.
4. Find $P(z < -1.9)$.
5. Find $P(-0.35 < z < 1.46)$.
6. Find the value of z^* such that $P(z < z^*) = 0.53$.
7. Find the value of z^* such that $P(z > z^*) = 0.27$.
8. Find the value of z^* such that $P(z < z^*) = 0.03$.

Few normal random variables actually have a standard normal distribution. However, any normal random variable can be transformed to a standard normal, and any standard normal random variable can be transformed to a normal random variable with any mean μ and standard deviation σ. Specifically, if $X \sim N(\mu,\sigma))$, $z = \dfrac{x - \mu}{\sigma} \sim N(0,1)$. Further, if $z \sim N(0,1)$, then $X = \mu + \sigma z \sim N(\mu,\sigma)$. Using these relationships, we can find probabilities and extreme values for any normal random variable using the z-table. When doing this, it is important to do all calculations carefully.

Example

Let $X \sim N(10,5)$. Find $P(X < 15)$.

Solution

$$P(X < 15) = P\left(\frac{X - \mu}{\sigma} < \frac{15 - 10}{5}\right)$$
$$= P(z < 1)$$
$$= 0.8413.$$

Notice that inside the parentheses, we had to transform both the X and the 15 to avoid changing the inequality. When working with X, we used symbols, and we used numbers when working with 15. However, we used the numbers that were associated with each symbol. Once we have the expression in terms of z, then the problem is equivalent to the earlier ones we worked. See Figure 11.15.

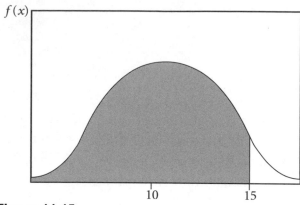

$f(x)$

Figure 11.15

Example

Let X, $N(5,2)$. Find X^* such that $P(X > X^*) = 0.05$.

Solution

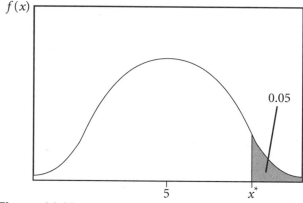

$f(x)$

0.05

5   x^*

Figure 11.16

First, we find z^* such that $P(z > z^*) = 0.05$. From our earlier work, we know that $z^* = 1.645$. Then $X^* = \mu \ 1 \ \sigma z^* = +2(1.645) = 8.29$.

▶ Practice

9. Let $X \sim N(11,8)$. Find $P(X > 17)$.

10. Let $X \sim N(8,3)$. Find X^* such that $P(X < X^*) = 0.27$.

▶ In Short

We have discussed two continuous distributions: the uniform and the normal. When every value in an interval is equally likely to occur, then we have a uniform distribution. The normal distribution is the most commonly used continuous distribution. Probabilities associated with a normal random variable must be found by using tables, calculators, or computers. When using tables, it is possible to use only one table. By tradition, the probabilities of a standard normal distribution are tabulated. Probabilities for other normal random variables are found by transforming the problem to one on the standard normal.

LESSON 12

Sampling Distributions and the *t*-Distribution

LESSON SUMMARY

The distribution of a random sample from a population is called the sample distribution, as described in Lesson 3. Using sample values, we can obtain estimates of the population parameters, such as the mean, the standard deviation, or a proportion. If we take another sample, it is very likely that the estimates from the second sample will differ from those of the first. Fortunately, this variation among estimates from different samples is predictable. The key to understanding this variation lies in gaining an understanding of sampling distributions (yes, these are different from sample distributions). Intuitively, it seems that as the sample size increases, we should do *better*. Sampling distributions in this lesson and the Law of Large Numbers in the next give us insight into what is meant by *better*.

▶ Sampling Distributions Defined

In Lesson 3, we learned that a parameter is a summary measure of a population, and a statistic is a summary measure of a sample. A statistic is some function of sample values that does not involve any unknown quantities (such as parameters). An important statistical idea is that parameters are fixed, but generally unknown, and that statistics are known from the sample, but vary. It is critical to always keep in mind whether we are thinking about a parameter or a statistic.

Suppose we are interested in the mean arm span of females attending college in the United States. (The arm span is the distance from the tip of one middle finger to the tip of the other middle finger when the arms are fully extended to the sides and perpendicular to the body.) Ten different people independently estimate the mean arm span by taking a random sample of 20 U.S. college females and finding the sample mean of the arm spans. It is very likely that this will lead to ten different estimates of the population mean arm span. Because each person selected a different sample, the arm spans of the people within each sample, and thus the sample means, will tend to be different. We have ten observations from the sampling distribution of the sample mean, one from each sample. The *sampling distribution* of a statistic is the distribution of possible values of that statistic for repeated samples of the same size from a population. That is, the statistic, the sample mean in our example, is a random variable, and the sampling distribution is the distribution for that random variable. Fortunately, we know quite a bit about the sampling distributions of the statistics that we will be most interested in.

Example

For each of the following circumstances, explain whether the quantity in bold is a parameter or a statistic.

1. For a sample of 20 married couples, there was a difference of **2.5** inches in the mean heights of the husbands and wives.
2. The Census Bureau reported that, based on the 2000 Census, the median age of residents of Oklahoma was **35.3** years.

Solution

1. The 2.5 inches is a statistic because it is based on a sample from the population of all married couples.

2. The 35.3 years is a parameter. A census occurs when every unit in the population is contacted. The U.S. Census occurs every ten years and attempts to contact everyone in the United States.

Example

For each of the following, specify whether $\overline{X}$ or μ is the correct statistical notation for each mean described.

1. A university administrator determines that the mean GPA of all students at the school is 2.4.
2. The mean amount of money spent on groceries for one week was $80.40 for a random sample of 35 single adult men.

Solution

1. Because all members of the population of interest (all students at the university) were included in finding the mean, the appropriate statistical notation is μ.
2. Here, we have the sample mean $\overline{X}$, which is an estimate of the mean amount spent by all single adult men (the population).

▶ Practice

For each of the following circumstances, explain whether the quantity in bold is a parameter or a statistic.

1. The Department of Transportation used a radar gun to measure the speed of a random sample of vehicles driving on a specific five-mile stretch of interstate from 8:00 to 8:30 A.M. The average speed of these vehicles was **63** miles per hour.
2. The average age of all residents at a nursing home is **85**.

For each of the following, specify whether $\overline{X}$ or μ is the correct statistical notation for each mean described.

3. The mean salary of teachers at a high school is $35,000 a year.

4. For a random sample of 35 college students, the average time per week they spent studying was 7.8 hours.

▶ Sampling Distribution of the Sample Mean

Consider again measuring the arm span of 20 randomly selected college females. Suppose arm spans of college females are normally distributed with a mean of 65.5 inches and a standard deviation of 2.5 inches. Thus, 68% of the population values will be between 63 and 68 inches, and 95% will be between 60.5 and 70.5 inches.

Suppose we take repeated samples of size 20. The samples, and thus the sample means, tend to vary with sample. Following are three samples that could have been drawn, where all measurements are in inches.

Sample 1: 67.7, 62.3, 66.6, 67.0, 62.3, 66.1, 61.2, 65.8, 67.0, 64.6, 60.4, 64.8, 66.7, 63.9, 65.2, 68.9, 63.1, 66.0, 67.3, 71.1

Sample 2: 67.5, 66.3, 66.7, 64.8, 66.5, 64.6, 68.5, 62.5, 64.9, 64.1, 68.9, 68.2, 62.9, 63.3, 64.5, 66.2, 66.4, 65.7, 66.8, 64.7

Sample 3: 63.9, 65.1, 68.7, 67.2, 60.2, 63.4, 63.9, 63.6, 68.0, 65.5, 65.6, 62.6, 63.9, 68.6, 65.6, 63.5, 66.3, 59.9, 67.9, 68.0

The sample means for samples 1, 2, and 3 are 65.4, 65.7, and 65.07 inches, respectively. None is exactly equal to the population mean (65.5 inches) or to each other, but all were fairly close.

This process of selecting samples of size 20 was repeated 10,000 times using computer simulation, and the histogram of the simulated sampling distribution of sample means is presented in Figure 12.1. In addition, the normal distribution that best fits the data is superimposed on the histogram. Notice that the normal curve provides a good representation of the histogram. The sample mean of the 10,000 sample means (the average of the sample means from all 10,000 samples of size 20) is 65.02, very close to the population mean of 65.5, and the sample standard deviation is 0.559. In addition, almost all of the values, not just 68% of them, are between 63 and 68 inches.

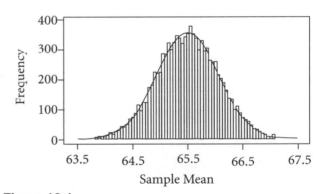

Figure 12.1

The sample standard deviation of the sample means is much smaller than the population standard deviation. Why? The population standard deviation is a measure of the spread in the arm span measurements of college females; it is a measure of the spread in the distribution of the random variable X, the arm span of a college female measured in inches. The sample standard deviation of the sample means based on samples of size 20 is a measure of the spread of the sample means, not the individuals; it is a measure of the spread in the sampling distribution of the random variable $\overline{X}$, the sample mean of the arm spans of 20 college females measured in inches.

What would happen if the sample size changed? To find out, we simulated 10,000 samples of size 50 from the same normal distribution with a mean of 65.5 and a standard deviation of 2.5. The sample mean was computed for each sample, and a histogram of this simulated sampling distribution is shown in Figure 12.2. Although the two graphs look similar at first glance, notice that the range of the sample means is quite different. For this set of 10,000 sample means, the average of the sample means is 65.494, and the sample standard deviation is 0.353, which is substantially less than the sample standard deviation of the sample means when the sample size was 20.

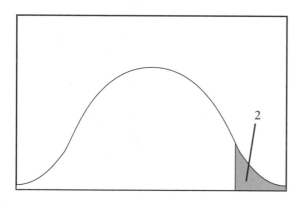

Figure 12.2

If a random sample of size n is taken from a normal distribution with mean μ and a standard deviation σ, the sampling distribution of the sample mean $\overline{X}$ is normal with a mean μ and a standard deviation $\frac{\sigma}{\sqrt{n}}$. That is, the sampling distribution of $\overline{X}$ is centered at the same value as the population distribution, but it has less spread. The spread in the sampling distribution, measured by $\frac{\sigma}{\sqrt{n}}$, decreases as the sample size increases. When we simulated the sampling distribution by generating 10,000 samples using

samples of size $n = 20$, the sample standard deviation of the sampling distribution of $\overline{X}$ was 0.559. We believe that it should be $\frac{\sigma}{\sqrt{n}} = \frac{2.5}{\sqrt{20}} = 0.55901$. (Here, we carried more decimal places than usual to aid the comparison.) These are really close! What about when $n = 50$? We have $\frac{\sigma}{\sqrt{n}} = \frac{2.5}{\sqrt{50}} = 0.35355$. Again, there is very good agreement with the simulated sampling distribution! This does not prove the statements are true; such a proof requires methods beyond this book. It does help us feel comfortable that the formula works.

Notice how closely the average of the 10,000 $\overline{X}$ values was to the population mean μ for samples of size $n = 20$ and $n = 50$, just as was predicted. The fact that the mean of the sampling distribution of the sample means is equal to the population mean indicates that the sample mean $\overline{X}$ is an unbiased estimator of the population mean μ. In general, an *unbiased statistic* is a statistic with mean value equal to the value of the population characteristic being estimated. We usually want to use an unbiased statistic to estimate a population parameter of interest.

If $\overline{X}$ has a normal distribution with mean μ and standard deviation $\frac{\sigma}{\sqrt{n}}$, by the properties of the normal distribution, we know that $z = \dfrac{\overline{X} - \mu}{\dfrac{\sigma}{\sqrt{n}}}$ has a standard normal distribution. That is, the sampling distribution of z is a standard normal. Thus, we can use the standard normal tables to find probabilities that the sample mean falls in intervals of interest, such as the probability the sample mean exceeds a specified value or is between two values, just as we did in Lesson 11.

Example

In a large population of high school students, the number of hours spent studying during any given week is normally distributed with mean 4.5 hours and standard deviation 18.9 hours. Consider randomly selected samples of size $n = 100$ students.

1. What is the mean of the sampling distribution of the sample means?
2. What is the standard deviation of the sampling distribution of the sample means?
3. Use properties of the normal distribution to fill in the blanks in the following sentence: "For 68% of all randomly selected samples of size $n = 100$ students, the mean amount of time spent studying during a week will be between ____ and ____ hours."

Solution

1. The mean of the sampling distribution of the sample means is 4.5 hours, the same as the population mean.
2. The standard deviation of the sampling distribution of the sample means is
$$\frac{\sigma}{\sqrt{n}} = \frac{18.9}{\sqrt{100}} = 1.89 \text{ hours.}$$
3. Because the population of the numbers of hours the high school students spent studying during the previous week is normally distributed, the distribution of the sample means is also normal. For a normal distribution, 68% of the population lies within one standard deviation of the mean. Thus, 68% of the sample means would lie within 1.89 hours (the standard deviation of the sampling distribution of sample means) of 4.5 hours (the mean of the sampling distribution of sample means). Thus, in repeated samples of $n = 100$, 68% of the samples will estimate the mean amount of time high school students spent studying during the previous week to be between 2.6 and 6.4 hours.

▶ Practice

In a large retirement community, the age of residents is normally distributed with mean 73 years and standard deviation 13 years. Consider randomly selected samples of size $n = 75$ residents.

5. What is the mean of the sampling distribution of the sample means?
6. What is the standard deviation of the sampling distribution of the sample means?
7. Use properties of the normal distribution to fill in the blanks in the following sentence: "For 95% of all randomly selected samples of size $n = 75$ residents, the mean age will be between ____ and ____ years."

▶ The *t*-Distribution

Again, assume that a random sample has been drawn from a normal distribution. The sampling distribution of $\overline{X}$ is normal. If we know the mean and standard deviation of the population, then we can answer many questions about $\overline{X}$ and the values it may assume. In practice, σ is almost never known. Most statisticians have never had a real-life problem in which the standard deviation was known! (The same is true for the mean, but we will address this issue later.) Assuming that σ was known, we know that $z = \frac{\overline{X} - \mu}{\frac{\sigma}{\sqrt{n}}}$ has a standard normal distribution. If we use the sample standard deviation instead of σ, then a different standardized random variable, denoted by t, results in the following:

$$t = \frac{\overline{X} - \mu}{\frac{s}{\sqrt{n}}}.$$

When working with z, only one quantity is varying with each sample, $\overline{X}$. For t, two quantities are varying, $\overline{X}$ and s. The value of s may not be very close to σ, especially for small values of n. Consequently, the distribution of t tends to be more variable than the distribution of z, especially for small n.

The *t*-distribution is centered at zero. It has one parameter, called the *degrees of freedom*, abbreviated as *df*. The degrees of freedom are usually a function of the sample size n, but the exact relationship between *df* and n depends on the type of problem. For this particular application we are considering, $t = \dfrac{\overline{X} - \mu}{\dfrac{s}{\sqrt{n}}}$.

The degrees of freedom are $(n - 1)$, one less than the sample size. The *t*-distribution is bell shaped and looks much like the standard normal, but it has thicker tails. The thicker tails reflect the increased variability for a *t*-distribution compared to the standard normal distribution. As the *df* increase, the tails become less thick and the distribution becomes more like the normal. The normal distribution and the *t*-distributions with 4 and 10 degrees of freedom are displayed in Figure 12.3.

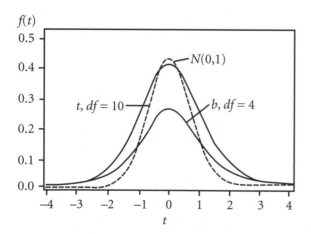

Figure 12.3

Like the normal, it is not easy to compute probabilities or to find specific values of t from the *t*-distribution. We must again rely on tabled values, calculators, or computers. For specified degrees of freedom, the tabulated t^* in the body of the table is chosen so that $P(t > t^*) = \alpha$. Notice that in the normal table we first used in the previous lesson, we worked with the left-tail probabilities. Here, we have the right-tail probabilities. Also, for the normal distribution, the probabilities were in the body of the table. For the *t*-distribution, the t^* values are in the body of the table (see Table 12.1).

Example

1. Find t^* such that $P(\text{t} > t^*) = 0.05$ when t has 8 degrees of freedom.
2. Find t^* such that $P(|t| > t^*) = 0.05$ when t has 12 degrees of freedom.
3. Find t^* such that $P(t < t^*) = 0.05$ when t has 16 degrees of freedom.

Solution

1. Here, we want the right-tail probability to be 0.05. Right-tail probabilities are presented in the table. To find the proper value, look at the row with 8 *df* and the column headed by 0.05. The two intersect at $t^* = 1.86$.
2. We should first recall from algebra that $P(|t| > t^*) = P(t > t^*) + P(t < -t^*)$. Because the *t*-distribution is symmetric, $P(t > t^*) = P(t < -t^*)$. Thus, we want to find t^* such that $P(t > t^*) = \dfrac{0.05}{2} = 0.025$. In the *t*-table, the row with 12 degrees of freedom intersects with the 0.025 column at $t^* = 2.179$. Thus, $P(t > 2.179) = 0.025$. By symmetry, we also have $P(t < -2.179) = 0.025$. This gives us that $P(|t| > t^*) = 0.05$.

Table 12.1 Probability α of exceeding the critical value

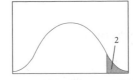

df	0.10	0.05	0.025	0.01	0.005	0.001
1.	3.078	6.314	12.706	31.821	63.657	318.313
2.	1.886	2.920	4.303	6.965	9.925	22.327
3.	1.638	2.353	3.182	4.541	5.841	10.215
4.	1.533	2.132	2.776	3.747	4.604	7.173
5.	1.476	2.015	2.571	3.365	4.032	5.893
6.	1.440	1.943	2.447	3.143	3.707	5.208
7.	1.415	1.895	2.365	2.998	3.499	4.782
8.	1.397	1.860	2.306	2.896	3.355	4.499
9.	1.383	1.833	2.262	2.821	3.250	4.296
10.	1.372	1.812	2.228	2.764	3.169	4.143
11.	1.363	1.796	2.201	2.718	3.106	4.024
12.	1.356	1.782	2.179	2.681	3.055	3.929
13.	1.350	1.771	2.160	2.650	3.012	3.852
14.	1.345	1.761	2.145	2.624	2.977	3.787
15.	1.341	1.753	2.131	2.602	2.947	3.733
16.	1.337	1.746	2.120	2.583	2.921	3.686
17.	1.333	1.740	2.110	2.567	2.898	3.646
18.	1.330	1.734	2.101	2.552	2.878	3.610
19.	1.328	1.729	2.093	2.539	2.861	3.579
20.	1.325	1.725	2.086	2.528	2.845	3.552
21.	1.323	1.721	2.080	2.518	2.831	3.527
22.	1.321	1.717	2.074	2.508	2.819	3.505
23.	1.319	1.714	2.069	2.500	2.807	3.485
24.	1.318	1.711	2.064	2.492	2.797	3.467
25.	1.316	1.708	2.060	2.485	2.787	3.450
26.	1.315	1.706	2.056	2.479	2.779	3.435
27.	1.314	1.703	2.052	2.473	2.771	3.421
28.	1.313	1.701	2.048	2.467	2.763	3.408
29.	1.311	1.699	2.045	2.462	2.756	3.396
30.	1.310	1.697	2.042	2.457	2.750	3.385
31.	1.309	1.696	2.040	2.453	2.744	3.375
32.	1.309	1.694	2.037	2.449	2.738	3.365
33.	1.308	1.692	2.035	2.445	2.733	3.356
34.	1.307	1.691	2.032	2.441	2.728	3.348
35.	1.306	1.690	2.030	2.438	2.724	3.340
36.	1.306	1.688	2.028	2.434	2.719	3.333
37.	1.305	1.687	2.026	2.431	2.715	3.326
38.	1.304	1.686	2.024	2.429	2.712	3.319
39.	1.304	1.685	2.023	2.426	2.708	3.313
40.	1.303	1.684	2.021	2.423	2.704	3.307
41.	1.303	1.683	2.020	2.421	2.701	3.301
42.	1.302	1.682	2.018	2.418	2.698	3.296
43.	1.302	1.681	2.017	2.416	2.695	3.291
44.	1.301	1.680	2.015	2.414	2.692	3.286
45.	1.301	1.679	2.014	2.412	2.690	3.281
46.	1.300	1.679	2.013	2.410	2.687	3.277
47.	1.300	1.678	2.012	2.408	2.685	3.273
48.	1.299	1.677	2.011	2.407	2.682	3.269
49.	1.299	1.677	2.010	2.405	2.680	3.265
50.	1.299	1.676	2.009	2.403	2.678	3.261

Table 12.1 *continued*

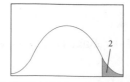

df	0.10	0.05	0.025	0.01	0.005	0.001
51.	1.298	1.675	2.008	2.402	2.676	3.258
52.	1.298	1.675	2.007	2.400	2.674	3.255
53.	1.298	1.674	2.006	2.399	2.672	3.251
54.	1.297	1.674	2.005	2.397	2.670	3.248
55.	1.297	1.673	2.004	2.396	2.668	3.245
56.	1.297	1.673	2.003	2.395	2.667	3.242
57.	1.297	1.672	2.002	2.394	2.665	3.239
58.	1.296	1.672	2.002	2.392	2.663	3.237
59.	1.296	1.671	2.001	2.391	2.662	3.234
60.	1.296	1.671	2.000	2.390	2.660	3.232
61.	1.296	1.670	2.000	2.389	2.659	3.229
62.	1.295	1.670	1.999	2.388	2.657	3.227
63.	1.295	1.669	1.998	2.387	2.656	3.225
64.	1.295	1.669	1.998	2.386	2.655	3.223
65.	1.295	1.669	1.997	2.385	2.654	3.220
66.	1.295	1.668	1.997	2.384	2.652	3.218
67.	1.294	1.668	1.996	2.383	2.651	3.216
68.	1.294	1.668	1.995	2.382	2.650	3.214
69.	1.294	1.667	1.995	2.382	2.649	3.213
70.	1.294	1.667	1.994	2.381	2.648	3.211
71.	1.294	1.667	1.994	2.380	2.647	3.209
72.	1.293	1.666	1.993	2.379	2.646	3.207
73.	1.293	1.666	1.993	2.379	2.645	3.206
74.	1.293	1.666	1.993	2.378	2.644	3.204
75.	1.293	1.665	1.992	2.377	2.643	3.202
76.	1.293	1.665	1.992	2.376	2.642	3.201
77.	1.293	1.665	1.991	2.376	2.641	3.199
78.	1.292	1.665	1.991	2.375	2.640	3.198
79.	1.292	1.664	1.990	2.374	2.640	3.197
80.	1.292	1.664	1.990	2.374	2.639	3.195
81.	1.292	1.664	1.990	2.373	2.638	3.194
82.	1.292	1.664	1.989	2.373	2.637	3.193
83.	1.292	1.663	1.989	2.372	2.636	3.191
84.	1.292	1.663	1.989	2.372	2.636	3.190
85.	1.292	1.663	1.988	2.371	2.635	3.189
86.	1.291	1.663	1.988	2.370	2.634	3.188
87.	1.291	1.663	1.988	2.370	2.634	3.187
88.	1.291	1.662	1.987	2.369	2.633	3.185
89.	1.291	1.662	1.987	2.369	2.632	3.184
90.	1.291	1.662	1.987	2.368	2.632	3.183
91.	1.291	1.662	1.986	2.368	2.631	3.182
92.	1.291	1.662	1.986	2.368	2.630	3.181
93.	1.291	1.661	1.986	2.367	2.630	3.180
94.	1.291	1.661	1.986	2.367	2.629	3.179
95.	1.291	1.661	1.985	2.366	2.629	3.178
96.	1.290	1.661	1.985	2.366	2.628	3.177
97.	1.290	1.661	1.985	2.365	2.627	3.176
98.	1.290	1.661	1.984	2.365	2.627	3.175
99.	1.290	1.660	1.984	2.365	2.626	3.175
00.	1.290	1.660	1.984	2.364	2.626	3.174
∞	1.282	1.645	1.960	2.326	2.576	3.090

3. This time, we are looking for a left-tail probability. We begin by finding the t^* that would provide the same size right-tail probability; that is, find t^* such that $P(t > t^*) = 0.05$. At the intersection of the row for 16 degrees of freedom and the 0.05 column, we find 1.746. By symmetry, the left-tail probability would be 0.05 for $t^* = -1.746$. Therefore, here $t^* = -1.746$.

▶ Practice

8. Find t^* such that $P(t > t^*) = 0.01$ when t has 27 degrees of freedom.

9. Find t^* such that $P(|t| > t^*) = 0.01$ when t has 23 degrees of freedom.

10. Find t^* such that $P(t < t^*) = 0.01$ when t has 20 degrees of freedom.

▶ Standard Errors

Throughout this lesson, we have talked about the "sample standard deviation of the sampling distribution of the sample mean." We have done this to emphasize that we are trying to estimate the standard deviation, not of the original population, but of the conceptual population of the sample means that could be derived from repeatedly taking random samples of size n and finding $\overline{X}$ for each. Because the variability in $\overline{X}$, $\widehat{p}$, and other estimators is so important, the term *standard error* is used to represent this idea. That is, the *standard error* of a statistic is the estimated standard deviation of the statistic. Thus, the "sample standard deviation of the sampling distribution of the sample mean" may be simply stated as the "standard error of $\overline{X}$." The standard error of the sample proportion $\widehat{p}$ is $\sqrt{\dfrac{\widehat{p}(1 - \widehat{p})}{n}}$. For a sample mean, the standard deviation of $\overline{X}$ is $\dfrac{\sigma}{\sqrt{n}}$ if σ is known so there would be no need to estimate that quantity using the standard error. However, in practice, one rarely, if ever, knows σ, so this case will not be considered further. If the mean is unknown, the standard deviation is generally unknown so the sample standard deviation is used to estimate it. Thus, when σ is unknown, the standard error of $\overline{X}$ is $\dfrac{s}{\sqrt{n}}$. Notice that as the sample size increases, the standard error decreases.

▶ In Short

Each sample produces a sample mean. If we selected another sample, the observations in the sample, and thus the value of the sample mean, are likely to change. When sampling from a normal distribution, the sampling distribution of the sample mean is also normal with a mean equal to the population mean and a standard deviation equal to the population standard deviation divided by the square root of n. The sample mean can be standardized using $z = \dfrac{\overline{X} - \mu}{\dfrac{\sigma}{\sqrt{n}}}$ if the population standard deviation is known and $t = \dfrac{\overline{X} - \mu}{\dfrac{s}{\sqrt{n}}}$. Because the sample standard deviation of the sampling distribution of the sample mean is frequently used, it is simply referred to as the standard error of $\overline{X}$.

13 ▶ The Law of Large Numbers and the Central Limit Theorem

LESSON SUMMARY

If a random sample of size n is taken from a population having a normal distribution with a mean μ and a standard deviation σ, then the sampling distribution of the sample mean $\overline{X}$ is also normal with a mean of μ and a standard deviation of $\frac{\sigma}{\sqrt{n}}$. What happens if the population distribution is not normal? Do we really get a better estimate of the population mean, or other parameters for that matter, if we take a larger sample? The Law of Large Numbers and the Central Limit Theorem will help us answer these questions.

▶ Law of Large Numbers

We learned in the last lesson that if a random sample is taken from a normal distribution, then the sampling distribution of the sample mean $\overline{X}$ is also normal. What happens when the population distribution is not normal? If a random sample of size n is selected from any distribution with mean μ and standard deviation σ, the sampling distribution of the sample mean $\overline{X}$ has mean μ and standard deviation $\frac{\sigma}{\sqrt{n}}$. This is true no matter what the form of the population distribution.

The standard deviation of the sampling distribution of $\overline{X}$ decreases as the sample size increases, and the sampling distribution of $\overline{X}$ is centered on the population mean. Thus, we know that, at least in some sense, the sample mean is getting closer to the population mean as the sample size increases. However, the Law of Large Numbers tells us even more. The *Law of Large Numbers* states that, provided the sample size is

Provided the sample size is large enough, the sample mean $\overline{X}$ will be "close" to the population mean μ with a specified level of probability, regardless of how small a difference "close" is defined to be.

large enough, the sample mean $\overline{X}$ will be "close" to the population mean μ with a specified level of probability, regardless of how small a difference "close" is defined to be. Unless we observe every member of the population, we can *never* be sure that the sample mean equals the population mean. However, in practice, we can be assured that, for any specific population, $\overline{X}$ is tending to get closer to μ as the sample size increases. In the previous lesson, we talked about the sample mean being a "better" estimate of μ as n increases. Here, we are saying that, as the sample size increases, we are doing better because the sample mean has an increased probability of being close to the population mean. Though we will focus on the mean, the Law of Large Numbers applies to other estimators, such as variance, as well.

▶ Central Limit Theorem

We now know that the sampling distribution of the sample mean $\overline{X}$ has a mean of μ and a standard deviation of $\frac{\sigma}{\sqrt{n}}$ no matter what form the population distribution has as long as the population has a finite mean and standard deviation. However, this alone is not enough for us to know what the shape of the sampling distribution is. Surprisingly, if the sample size is sufficiently large, the sampling distribution of $\overline{X}$ is approximately normal! This follows from a basic statistical result, the Central Limit Theorem. The *Central Limit Theorem* states that, if a sufficiently large random sample of size n is selected from a population with finite mean μ and finite standard deviation σ,

the sampling distribution of the sample mean $\overline{X}$ is approximately normal with mean μ and standard deviation of $\frac{\sigma}{\sqrt{n}}$. That is, as long as the sample is large enough, the normal distribution serves as a good model for the sampling distribution of $\overline{X}$ and it does not matter whether the population is normal or non-normal or even discrete or continuous. We will illustrate this with a poll.

Suppose a poll is conducted to determine what proportion of the registered voters in a large city thinks that the sales tax should be increased so that more recreational facilities, such as public parks and swimming pools, can be developed. One hundred registered voters in the community could be surveyed. A response supporting the increased sales tax is recorded as $X = 1$, and a response against the increased sales tax is recorded as $X = 0$. Then the sample mean $\overline{X}$ is the sample proportion, which we will denote as $\widehat{p}$. That is, $\overline{X}$ (or equivalently $\widehat{p}$) is the sample mean that estimates the population proportion p, the proportion of the registered voters in the large city who favor the increase in sales tax. Because each registered voter polled either agrees or disagrees (only two possible responses), the random selection of a single registered voter whose response is recorded is equivalent to a Bernoulli trial. Suppose that 68% of the registered voters support the increased sales tax. (Of course, we do not *really* know what proportion favors the sales tax. If we did, there would be no need for the survey.) A graph of the probabilities associated with the responses is shown in Figure 13.1. The population is discrete and not symmetric; it is certainly not normal!

If a sufficiently large random sample of size n is selected from a population with finite mean μ and finite standard deviation σ, the sampling distribution of the sample mean $\overline{X}$ is approximately normal with mean μ and standard deviation $\dfrac{\sigma}{\sqrt{n}}$.

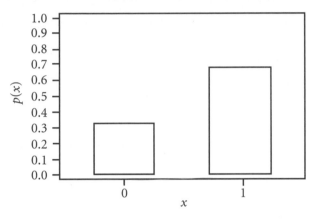

Figure 13.1

Samples of size 100 were drawn from a Bernoulli distribution with $p = 0.68$, and the sample mean $\widehat{p} = \overline{X}$ calculated for each sample. Assuming samples of size 100 are large enough for the Central Limit Theorem to apply, we expect the sampling distribution of $\widehat{p}$ to be approximately normal. Further, because the mean and variance of a Bernoulli random variable are p and $p(1 - p)$, respectively, we anticipate the standard deviation of the sampling distribution of $\widehat{p}$ to be $p = 0.68$ and

$$\frac{\sqrt{p(1 - p)}}{\sqrt{n}} = \frac{\sqrt{0.68(1 - 0.32)}}{\sqrt{100}} = 0.0466,$$

respectively.

A histogram of the simulated sampling distribution based on 10,000 samples of size 100 from a Bernoulli distribution with $p = 0.68$ is shown in

Figure 13.2. The bell-shaped appearance suggests that the distribution is approximately normal as predicted by the Central Limit Theorem. The average of the 10,000 $\widehat{p}$ values is 0.6807, and the sample standard deviation of the 10,000 $\widehat{p}$ values is 0.04668. These are very close to the values of p and $\dfrac{\sqrt{p(1 - p)}}{\sqrt{n}}$ that we anticipated based on the properties of the sampling distribution of $\overline{X}$. (Remember, as long as the sample is randomly selected, these properties hold regardless of the shape of the population distribution as long as it has a finite mean and a finite variance.)

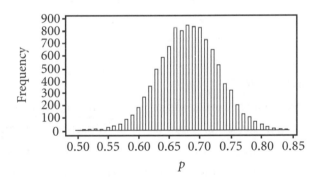

Figure 13.2

From the previous polling example, we can conclude that when $p = 0.68$, it is reasonable to assume that the sampling distribution of the sample proportion $\widehat{p}$ is approximately normal when the sample size is 100. Would a sample size of 50 have been large enough? What about 10? Does it depend on p? Such questions have been explored, and guidelines have been developed. For proportions, the rules apply to two cases: (1) a population has a fixed proportion who has a certain trait, opinion, etc. or (2) a repeatable experiment in which a certain outcome will occur with a constant probability p. Suppose we take a random sample of size n in the first case and repeat the experiment n times in the second case. If $np \geq 10$ and $np(1 - p) \geq 10$, then the sample size n is large enough for the Central Limit Theorem to apply. Of course, p is usually unknown, so we check that these conditions hold using $\widehat{p}$.

The population distribution could be some other discrete or continuous distribution. Suppose we take a random sample of size n from one of these. The Central Limit Theorem tells us that if n is large enough, the sampling distribution of $\overline{X}$ is approximately normal. An arbitrary rule is often given that the Central Limit Theorem applies for $n \geq 30$. Although we will tend to use this rule, if the population distribution is highly skewed or if there are extreme outliers, a larger sample would be better.

Example

Based on a random sample of 4,252 men for the 1988 National Survey of Families and Households, it was reported that men spent a mean of 18.1 hours per week doing housework. Suppose the standard deviation was known to be 12.9 hours.

1. Assuming a normal distribution and using the information given, sketch the approximate distribution of the number of hours a randomly selected man spent doing housework in 1988.
2. Based on the graph in the previous question, explain why the population distribution is very unlikely to be normal and why it is most likely skewed to the right.

3. The researcher believes that 18.1 hours per week is close to the mean time men spent doing housework in 1988. Provide a justification for this reasoning.
4. What is the approximate sampling distribution of the mean number of hours men spent on housework in 1988 based on samples of size 4,252?

Solution

1. Examine the graph in Figure 13.3.

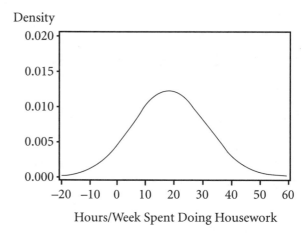

Density

Hours/Week Spent Doing Housework

Figure 13.3

2. For a normal distribution, 95% of the population is within two standard deviations of the mean. If we assume for the moment that the sample mean is a good estimate of the population mean, 95% of the men would spend between -7.7 and $+43.9$ hours on housework. It is impossible for anyone to work a negative number of hours; the minimum number of hours that could be recorded is zero. Given the large sample size, it is unlikely that the sample mean is such a poor estimate that all values would be nonnegative if we had the true population mean. Thus, the normal is not a reasonable

model for this distribution. To have only non-negative population values with a mean and standard deviation similar to those observed, the population distribution must be skewed to the right.

3. Because the sample size is large, the sample mean should be close to the population mean by the Law of Large Numbers.

4. The sample size of 4,252 is certainly large enough to invoke the Central Limit Theorem ($n \geq 30$). Therefore, the sample mean has an approximate normal distribution with an estimated mean of 18.1 hours and a standard deviation of $\dfrac{\sigma}{\sqrt{n}} = \dfrac{12.9}{\sqrt{4,252}} = 0.20$ hours.

Example

A female high school student decided that she wanted to date only males who went to her school and had brown eyes. To estimate what proportion of the males in her school would meet these criteria, she randomly selected 20 males from the school. Of these, 12 had brown eyes. Explain why the Central Limit Theorem does not apply.

Solution

The proportion of males in the school with brown eyes is estimated to be $\widehat{p} = \dfrac{12}{20} = 0.60$. To invoke the Central Limit Theorem, we must have $n\widehat{p} \geq 10$ and $n(1 - \widehat{p}) \geq 10$. Here, $n\widehat{p} = 20 \times 0.60 = 12 > 10$, but $n(1 - \widehat{p}) = 20(1 - 0.6) = 8 < 10$. Only one of the two necessary conditions is met, so it would not be appropriate to apply the Central Limit Theorem here.

▶ Practice

The receptionist believes that the doctor he works for requires the patients to wait too long. He randomly selects 100 patients over a six-week period and records the time each waits before being taken back to see the doctor. The receptionist finds that the average wait time was 20 minutes, and the sample standard deviation was 15 minutes.

1. Assuming a normal distribution and using the information given, sketch the approximate distribution of the number of minutes a randomly selected patient spent waiting to see the doctor.

2. Based on the graph in the previous practice question, explain why the population distribution is very unlikely to be normal and why it is most likely skewed to the right.

3. The receptionist believes that 20 minutes is close to the mean time patients spent waiting to see the doctor. Do you agree? Provide a justification for your response.

4. Based on samples of size 100, what is the approximate sampling distribution of the mean number of minutes patients spent waiting to see the doctor?

▶ In Short

The Law of Large Numbers assures us that the sample mean is getting closer to the population mean as the sample size increases. As long as a large enough sample is taken from a distribution with mean μ and standard deviation σ, the Central Limit Theorem assures us that $\overline{X}$ has an approximate normal distribution with mean μ and standard deviation $\frac{\sigma}{\sqrt{n}}$. We will repeatedly use this fact to draw inferences about populations.

Sample Surveys

LESSON SUMMARY

The results of surveys are presented almost daily in newspapers, over the radio, and on television. From surveys, the proportion p of the population with a certain trait or opinion can be estimated. In fact, if the sample size is 1,500, we can be almost sure that our estimate $\widehat{p}$ is within 0.03 of the population proportion. Remarkably, being able to estimate the population proportion with this precision does not depend on the size of the population. A sample of 1,500 people is sufficient whether we are drawing inference to the people living in a particular state, to the people living within the United States, or to the people living on Earth, provided the sample is taken properly. Taking a proper sample is challenging. In this lesson, we will learn more about conducting sample surveys.

► Margin of Error

From June 24 through 26, 2005, the Gallup Organization contacted 1,009 adults nationally and asked them, "How patriotic are you? Would you say—extremely patriotic, very patriotic, somewhat patriotic, or not especially patriotic?" Of the respondents, 72% said "extremely or very patriotic." Thus, $\widehat{p} = 0.72$ is the estimate of the proportion p of adults in the United States who would state they are extremely or very patriotic. The sample proportion is a point estimate of the population proportion. A *point estimate* of a population parameter is a single number that is based on sample data and represents a plausible value of the parameter.

The Gallup Organization also reported that there was a ± 3 percentage point margin of error associated with the survey. The margin of error provided by this and other media descriptions of survey results has two important characteristics. First, the difference between the sample proportion $\hat{p}$ and the population proportion p is less than the margin of error about 95% of the time; that is, for about 19 of every 20 random samples of the same size from the same population, the sample proportion will be within the margin of error of the population proportion. Second, the sample proportion will differ from the population proportion by more than the margin of error about 5% of the time; that is, for about one in every 20 samples of the same size from the same population, the difference in the sample proportion and the population proportion will be greater than the margin of error.

The margin of error can be used to obtain an interval of plausible values for the parameter of interest. For the survey on patriotism, the point estimate was 0.72, and the margin of error was 0.03. Thus, the interval of plausible values based on this sample is 0.69 to 0.75.

Example

A sample of high school students was randomly selected from a very large city. Each student was asked, "Are you employed either part time or full time during the school year?" Of those sampled, 38% reported that they had a part-time or a full-time job during the school year. The margin of error was reported to be 5%. Give a point estimate and an interval of reasonable values for the proportion of this city's high school students having employment that, with 95% certainty, includes the true proportion.

Solution

The point estimate of the proportion of the high school students in this city who are employed, either part time or full time, is 0.38. An interval of plausible values for this proportion is between $0.38 - 0.05 = 0.33$ and $0.38 + 0.05 = 0.43$.

▶ Practice

1. A group of students was randomly selected from a large university and asked if they thought that their university's campus was safe or unsafe. Of those sampled, 83% felt that their campus was safe. The margin of error was reported to be 4%. Give a point estimate and an interval of reasonable values for the proportion of this university's students who think that their campus is safe that, with 95% certainty, includes the true proportion.

▶ Census versus Sample Survey

In a census, every unit in the population is included in the sample. This is the only way to determine a parameter exactly. If our goal is to determine a parameter's value, why do we usually sample and not take a census? There are various reasons that we must, or want to, sample instead of taking a census.

It may not be feasible to take a census. When a nurse draws blood for a test, you certainly want her to be satisfied with a sample and not to take all of your blood as a census would require. A manufacturer who takes a census to determine the mean lifetime of the batteries the company produces will have nothing left to sell!

Many times, a census takes too long to complete. Suppose we want to know what proportion of the cotton plants in a 160-acre field has at least one insect on them. (The number of plants per acre can vary from 30,000 to 58,000.) It would take days to check each plant. By then, the plants first inspected that had insects may or may not still have insects on them, and the plants inspected early that did not have insects might now have insects on them. The U.S. Census, which is completed every ten years, takes years to plan and more time to compile the results after the data are collected; it would not be feasible to census the U.S. population each year.

A census is often not as accurate as a sample survey. A small group of interviewers can be trained more easily than can a large one. Finding a small number of nonrespondents is a much more manageable task than finding a large number of nonrespondents. For the U.S. Census, it is difficult to actually count all citizens. Some do not have a home; others do not want to be counted. Various techniques have been used to count these people. This has led some to argue that a more accurate count of the U.S. population would be obtained if it were estimated from a sample; others disagree.

▶ Simple Random Samples

Earlier, we noted that the sample proportion will be within the margin of error of the population mean *provided* that the sample was properly taken. Before describing some of the methods that can be used to select samples properly, we need to think more carefully about some of the elements of sampling.

In a sample survey, the target population is the set of units that is of interest. The sampled population is the set of units from which the sample is selected. Although we want the target and sampled populations to be the same, this is rarely the case. As an illustration, suppose the target population is every household in the United States. If a telephone survey is conducted

using the white pages from phone books across the nation, only households with telephone numbers in the white pages are part of the sampled population. Households without a telephone or with unlisted numbers are not part of the sampled population. The sample frame is a list of all units from which the sample is drawn; it is a list of the units in the sampled population.

Generally, the purpose of a sample survey is to draw inferences about some population characteristic(s). For a relatively small sample to accurately reflect the characteristics of a large population, the sample cannot be drawn haphazardly. Proper sampling methods, specifically, probability sampling plans, must be used. A *probability sampling plan* is one in which every unit of the sampled population has a known probability of being included in the sample. In Lesson 2, we learned that a simple random sample is one in which every set of units of size *n* in the sampled population has an equal chance of being selected; a random sample is a probability sample.

Suppose we want to take a simple random sample of size 30 from the people who have donated funds to the local public radio station within the past year. Working with the station, we could write each contributor's name on a slip of paper, place it in a bowl, mix the pieces of paper thoroughly, and draw out 30 slips. The names on the 30 slips of paper constitute the people in the sample. This approach becomes impractical as the population of interest becomes large. Writing the names of all residents of a city, much less a state or nation, on slips of paper would take a prohibitive amount of time. Instead, the sample frame (list of names) is usually generated from one or more sources, such as tax rolls or residential addresses, and the computer is used to make selections from the list in a manner that permits every listed unit (person) to have an equal chance of being chosen. Those units (people) selected by the computer constitute the simple random sample.

Generating the sample frame is a major challenge, especially if the population is large and/or geographically dispersed. The resources available to create the frame may not be sufficient. Sometimes, even if they are sufficient, it is impossible to create the sample frame, at least within the desired time frame. Other sampling methods have been developed as alternatives to simple random sampling. These tend to be more complicated both in selecting the sample and in obtaining parameter estimates from the sample. Depending on the circumstance, they may have some advantages over simple random sampling. We will consider four such methods: stratified random sampling, cluster sampling, systematic sampling, and multistage sampling.

▶ Stratified Random Sampling

Sometimes, the population has natural groups, called *strata*. As an illustration, when estimating the literacy rate for the nation, estimates of the literacy rate for each state may be useful in assessing state and regional differences. A *stratified random sample* is one in which the population is first divided into groups (strata) and then a simple random sample is taken within each stratum. Estimates are made for each stratum and then combined to obtain the population estimate. Because the sizes of strata usually vary, a weighted average of the stratum estimates, with weights proportional to the strata sizes (not a simple average), is used to estimate the population parameter.

▶ Cluster Sampling

Although cluster sampling is often confused with stratified random sampling, it is very different. In cluster sampling, a population is divided into groups, called *clusters*, a random sample of clusters is selected, and only units in those clusters are measured. In most applications of stratified random sampling, the population is divided into a few large strata and a simple random sample is selected from each stratum. In contrast, in most applications of cluster sampling, the population is divided into many small clusters, a sample of clusters is randomly selected, and every unit in the cluster is measured.

Cluster sampling is often used because it is easier and more cost effective than other alternatives. For example, suppose we want to sample the households in a large city, using door-to-door interviews. It may be very expensive to construct a list of all households, select n addresses at random, and visit each selected household. A cluster sample in which blocks within the city are randomly selected and all households within each selected block are interviewed may be more cost effective. Once a block is selected, the interviewer can conduct several interviews before moving to the next block, reducing the time needed to obtain interviews from the same number of households. However, households within the same block may tend to be more alike than households in different blocks. This tendency of units in the same cluster to be more alike than units in different clusters must be addressed in the analysis. Such approaches are available in books on sampling.

▶ Systematic Sampling

Suppose that you have a sample frame consisting of a list of 5,000 names and want to draw a sample of 100. To use a systematic sampling plan, we would divide the list into 100 consecutive segments of size $\frac{5,000}{100} = 50$, choose a random point in the first segment, and include that unit in the study and every unit at the same point in all segments. Upon completion, the sample would consist of 100 units equally spaced throughout the list. Systematic sampling is also used in many natural resource studies. Here, a grid of points is randomly placed over the region. To randomize, one point, say a corner point, is randomly assigned a location within a small area, and the whole grid is set relative to the random placement of that point.

Systematic sampling can be a good alternative to simple random sampling. If the sample units are randomly listed in the sample frame, the systematic sample is usually treated as a simple random sample. Care must be taken as systematic random sampling could lead to biases. The potential biases associated with treating a systematic sample as a simple random sample when using a grid have been discussed in the natural resources literature.

▶ Multistage Sampling

Many large surveys use a combination of the methods we have discussed. As an illustration, a large national survey may first stratify by regions of the country. Within each regional stratum, we might then stratify by state. Within each state, we could stratify by urban, suburban, and rural areas. We could then randomly select communities within each of the urban, suburban, and rural strata. Finally, we could randomly select blocks or fixed areas within each selected community and interview everyone within that fixed area. A *multistage sampling* plan is one that combines methods as illustrated here.

▶ Random-Digit Dialing

Most of the national polling organizations and many of the government surveys in the United States now use a sampling plan called random-digit dialing. This method approximates a random sample of all households in the region of interest that have telephones. To initiate a random-digit dialing plan, the polling organization must first get a list of all telephone exchanges in the region of interest. A telephone exchange consists of the area code and the next three digits. Using the numbers listed in the white pages, the proportion of all households in the region with that specific exchange can be approximated. That proportion is used to determine the chance that the telephone exchange is randomly selected for inclusion in the sample. Next, the same process is followed to randomly select banks within each exchange. A telephone bank consists of the next two numbers. Finally, the last two digits are randomly selected from 00 to 99. Although the process is quite involved, it has been computerized, and random telephone numbers can be generated rapidly.

Once a telephone number has been generated, pollsters should make multiple attempts to reach someone at that household if no one responds initially. They may ask to speak to a male or an adult because females and children are more likely to answer the phone, potentially biasing the results because they are overrepresented in the sample.

Example

In each question, identify whether the sample is a stratified random sample or a cluster sample.

1. A large school system decided to survey sophomores within its high schools. Three schools were randomly selected, and all sophomores in those schools were included in the sample.

2. A professional society wants to improve its member services. It takes a simple random sample of current members and another simple random sample of members who have chosen not to renew. All people selected are surveyed.

Solution

1. This sampling plan leads to a cluster sample. The schools are the clusters, which are randomly selected. The sophomores within each school are the cluster members.

2. This sampling plan results in a stratified sample. There are two strata: current members and members who have not renewed their membership. A simple random sample is selected within each stratum.

▶ Practice

In each situation, identify whether the sample is a stratified random sample or cluster sample.

2. A large environmental organization holds a meeting every two years. A simple random sample of female members and another simple random sample of male members were selected. Each selected person was asked whether he or she enjoyed the speakers at this year's meeting.

3. Researchers were curious as to how often adults living in Florida exercised. Ten counties were randomly selected. All adults living in these counties were contacted to determine how often they exercise.

▶ In Short

Sample surveys are very useful for estimating the proportion of population units with a specific trait or opinion. Selecting a sample that allows us to make unbiased estimates of this proportion is challenging. Simple random sampling is appealing and leads to simple estimates, but selecting a true simple random sample may be physically impossible. For large surveys, stratified random samples, cluster samples, systematic samples, or a combination of these in a multistage sample may be easier to implement, more cost effective, and produce good estimates. Random-digit dialing approximates a simple random sample of all households with telephones in the region of interest.

Confidence Intervals for Proportions

LESSON SUMMARY

Studies are conducted and samples are drawn to learn more about one or more populations. If the form and parameters of the population distribution are known, there would be no need to sample. Sampling gives us information on the parameters of the distribution, but without a census, the population parameters cannot be determined exactly. The statistic estimating the parameter is rarely equal to the parameter. How close is the statistic to the parameter it is estimating? Can statements be made that an estimate is within a certain distance of the parameter with a known probability of the statement being correct? We will answer these questions during this lesson.

▶ Confidence Intervals for Proportions

Based on the Gallup poll on patriotism mentioned in the previous lesson, the proportion of U.S. adults who identify themselves as extremely or very patriotic is 0.72, and the margin of error was 0.03. Using this, we obtained an interval of values, from 0.69 to 0.75, that were plausible for the proportion of the U.S. adult population who characterize themselves as extremely or very patriotic. The interval 0.69 to 0.75 constituted a 95% confidence interval of the true population proportion. How was the margin of error computed? What do we do if we want some confidence level other than 95%? We will now outline the process of finding a confidence interval for a population proportion so that we can answer these questions.

Suppose the goal of a study is to estimate the population proportion p with $(1 - \alpha)100\%$ confidence. Provided that the sample size is sufficiently large (i.e., $np \geq 10$ and $n(1 - p) \geq 10$), a confidence interval of the form

$$\widehat{p} \pm z^* \sqrt{\frac{\widehat{p}(1 - \widehat{p})}{n}}$$

will have the desired level of confidence if z^* is chosen so that $P(-z^* < z < z^*) = 1 - \alpha$. To see this, we will have to put together several ideas from earlier lessons with the new concept of a confidence interval.

First, by the Central Limit Theorem, we know that, if the sample size n is sufficiently large, the sample proportion $\widehat{p}$ is approximately normally distributed with mean p and standard deviation

$\dfrac{\sigma}{\sqrt{n}} = \sqrt{\dfrac{p(1 - p)}{n}}$. Because p is unknown, the standard deviation of $\widehat{p}$ is unknown, but the standard error of $\widehat{p}$ $\sqrt{\dfrac{\widehat{p}(1 - \widehat{p})}{n}}$, provides an estimate of the standard deviation.

Using the properties of the normal distribution and again assuming the sample size is large enough, we can transform $\widehat{p}$ to an approximate standard normal random variable z by subtracting the mean and dividing by the standard error; that is:

$$z = \frac{\widehat{p} - p}{\sqrt{\dfrac{\widehat{p}(1 - \widehat{p})}{n}}}$$

Note that we have divided by the standard error of $\widehat{p}$ instead of the standard deviation of $\widehat{p}$. How large must the sample size be for the normal approximation to be adequate? The guidelines are the same as those we had for invoking the Central Limit Theorem in Lesson 13. If $np \geq 10$ and $n(1 - p) \geq 10$, the normal distribution provides a good approximation of the distribution of $\widehat{p}$. Because p is unknown, we use $\widehat{p}$ to check the conditions.

We also know that, for a standard normal distribution, we can find z^* such that a specified percentage of the population values are between $-z^*$ and z^*; the probability that a randomly selected value of z will be between $-z^*$ and z^* is the confidence level. Figure 15.1 illustrates the relationship in z^* and confidence level.

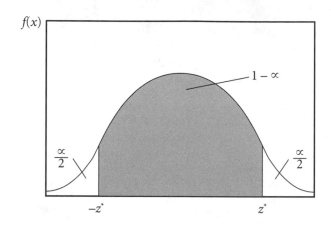

Figure 15.1

Table 15.1 provides z^* values for the most common levels of confidence.

Table 15-1 z-values

LEVEL OF CONFIDENCE	z*
90	1.645
95	1.96, often rounded to 2
98	2.33
99	2.58

Combining all of the above ideas, we have:

$$P\left(-z^* < \frac{\widehat{p} - p}{\sqrt{\dfrac{\widehat{p}(1 - \widehat{p})}{n}}} < z^*\right) = 1 - \alpha$$

Using algebra, we can rewrite the equation as:

$$P\left(\widehat{p} - z^*\sqrt{\frac{\widehat{p}(1 - \widehat{p})}{n}} < p\right.$$

$$\left. < \widehat{p} + z^*\sqrt{\frac{\widehat{p}(1 - \widehat{p})}{n}}\right) = 1 - \alpha$$

The limits on the inequality are the same as the confidence interval limits that we stated previously. Notice that the form of the confidence interval is *point estimate ± multiplier × standard error*.

For proportions, the point estimate is $\widehat{p}$, the multiplier is the value of z^* corresponding to the desired confidence level, and the standard error is

$\sqrt{\dfrac{\widehat{p}(1 - \widehat{p})}{n}}$. This general form will be seen again when we set confidence intervals on the mean.

Although it has been stated several times before, it is important to remember that the population proportion p is fixed. The confidence interval depends on the sample, and the limits of the confidence interval vary with the sample. To illustrate this, suppose we repeatedly draw samples of $n = 50$ from a population with $p = 0.6$. A confidence interval on p is found for each sample. The results of the confidence intervals from 100 of these samples are shown in Figure 15.2. The line segments represent the confidence intervals. Notice that four of the segments do not cross the vertical line at $p = 0.6$. That means 96 of the 100, or 96%, do include the population proportion 0.6. This is close to the predicted 95%. As the number of samples gets larger, the observed percentage will tend to get closer to the specified confidence level of 95%. Notice, the ends of the confidence intervals (the confidence limits) change with the sample, but the population proportion does not change.

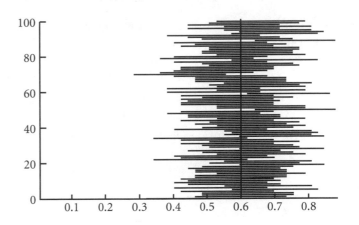

Figure 15.2

Example

A company wants to know what proportion of the bell pepper seeds it sells will germinate. One hundred seeds are randomly selected from the company's inventory. They are placed in ideal conditions for germination. After two weeks, 78 of the seeds had germinated. Set a 90% confidence interval for the proportion of seeds in inventory that would germinate under ideal conditions. Interpret the interval in the context of the problem.

Solution

First, we need to determine whether the conditions are satisfied for us to use a normal approximation to find the interval.

$n\widehat{p} = 100(0.78) = 78 > 10$
and $n(1 - \widehat{p}) = 100(1 - 0.78) = 12 > 10$

Because both $n\widehat{p}$ and $n(1 - \widehat{p})$ are greater than 10, we can use a large-sample confidence interval based on the normal distribution.

Second, we know that the standard error of $\widehat{p}$ is $\sqrt{\dfrac{0.78(1 - 0.78)}{10}}$. This would have been true even if the sample size had not been large enough to use a large-sample confidence interval.

Next, we need to find z^* such that $P(-z^* < z < z^*) = 0.90$. From the table of common z^* values provided earlier, $z^* = 1.645$ is the multiplier for a 90% confidence interval.

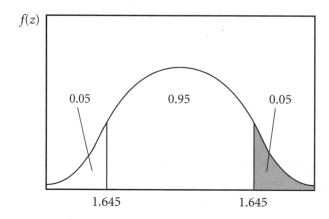

Figure 15.3

The limits of the confidence interval are then

$$0.78 \pm 1.645\sqrt{\dfrac{0.78(1 - 0.78)}{10}} \quad \text{or} \quad 0.78 \pm 0.22$$

The confidence interval is $(0.56, 1.00)$.

Interpretation: We are 90% confident that between 56% and 100% of the bell pepper seeds in this company's inventory would germinate under ideal conditions.

▶ Practice

A law school wants to know how successful its graduates are at passing the bar exam on the first try. Two hundred graduates are randomly selected and asked whether they passed the bar exam the first time they took it. Of the 200 graduates, 130 said they passed the exam on the first try.

1. Set a 98% confidence interval for the proportion of law student graduates at this university who passed the bar exam. Interpret the interval in the context of the problem.

2. Give a potential source of bias in this study.

▶ Sample Sizes, Confidence Level, and Length of Confidence Intervals

Recall that the general form of a confidence interval is *point estimate* $\pm$ *multiplier* $\times$ *standard error*. The length of the confidence interval is 2 $\times$ *multiplier* $\times$ *standard error*; the half-length of the confidence interval is *multiplier* $\times$ *standard error*. From Lesson 13, we know that as the sample size increases, the standard error decreases, so the length of the confidence interval decreases. Notice from the table of z^* values that the value of the multiplier increases as the confidence level increases, making the confidence interval longer. The fact that the confidence interval gets longer as the confidence level increases holds for other forms of intervals as well.

Example

In the last example, we set a 90% confidence interval on the proportion of seeds in inventory that would germinate under ideal conditions. Now set a 95% confidence interval on this same proportion. Compare the two intervals.

Solution

Changing the confidence level has no effect on whether or not the conditions for inference are satisfied. We still have

$n\widehat{p} = 100(0.78) = 78 > 10$ and
$n(1 - \widehat{p}) = 100(1 - 0.78) = 12 > 10$

so we can use the normal distribution to approximate the sampling distribution of $\widehat{p}$. Because the sample size has not changed, the standard error is the same. However, the multiplier z^* is different. We must have 0.025 probability in each tail instead of the 0.05 in each tail that corresponded to a 90% confidence interval. From the table, we have $z^* = 1.96$. Thus, the 95%

confidence interval is $0.78 \pm 1.96\sqrt{\dfrac{0.78(1 - 0.78)}{10}}$ or 0.78 ± 0.26 compared to the 90% confidence interval of 0.78 ± 0.22 found earlier. Clearly, the 95% confidence interval is wider than the 90% confidence interval. It makes sense that, as the interval increases in length, we become more confident that the interval will capture the true population proportion.

We should also note that the 95% confidence interval ranges from 0.52 to 1.04. However, p cannot be greater than 1. Therefore, rounding the upper limit to the largest admissible value, we would say that we are 95% confident that the proportion of bell pepper seeds in inventory that would sprout under optimal conditions is between 56 and 100%. Alternately, we might say with 95% confidence that at least 56% of the bell pepper seeds in inventory will sprout under optimal conditions.

▶ Practice

Consider the law school's study of the proportion of its graduates who pass the bar exam on the first try. Earlier, we had a sample of 200 and set a 98% confidence interval on the proportion of students who passed the bar exam on the first try.

3. Set a 99% confidence interval for the population proportion of graduates who pass the bar exam on the first try. Compare the two intervals.
4. Suppose that only 100 graduates were sampled and 65 reported passing the bar exam on the first try. (Note that the estimated proportion is the same for both samples.) Set a 99% confidence interval on the proportion of the school's graduates who pass the bar exam the first time. Compare the intervals.

▶ In Short

Confidence intervals provide a plausible set of values for the unknown population parameter of interest. Appealing either to normality or the Central Limit Theorem, confidence intervals have the form

point estimate $\pm$ multiplier $\times$ standard error.

For proportions, the point estimate is $\widehat{p}$, and the standard error is $\sqrt{\dfrac{\widehat{p}(1 - \widehat{p})}{n}}$. The multiplier z^* is chosen so that the interval has the desired level of confidence. Relationships exist among the length of the confidence interval, the sample size, and the level of confidence.

16 ▶ Hypothesis Testing for Proportions

LESSON SUMMARY

A confidence interval on the population proportion provides a set of plausible values for that proportion as we saw in the last lesson. If the proportion is hypothesized to be, say, 0.4, but the interval does not include 0.4, then it would be reasonable to reject that value for the population proportion. We will not always want to assess the validity of hypotheses using a confidence interval. In this lesson, the logic of statistical hypothesis testing and the application of this logic to proportions will be presented.

▶ Logic of Hypothesis Testing

To conduct a statistical test of hypotheses, we must first have two hypotheses: the null hypothesis and the alternative hypothesis. The *null hypothesis,* denoted by H_0, is a statement that nothing is happening. The specific null hypothesis varies depending on the problem. It could be that medication does not alter blood pressure, that no relationship exists between IQ and grades, or that hair grows at the same rate for females and males. The *alternative hypothesis,* denoted by H_a, is a statement that something is happening. As with the null hypothesis, the alternative hypothesis depends on the problem. It could be that medication changes blood pressure, that a relationship does exist between IQ and grades, or that hair grows at different rates for females and males.

For a moment, consider a trial. The null hypothesis at every trial is H_0: The defendant is not guilty. That is, she did not do whatever she is accused of. The alternative hypothesis is always H_a: The defendant is guilty. That is, she did commit the crime of which she is accused. In the U.S. judicial system, the jury is instructed that the null hypothesis of not guilty can be rejected in favor of the alternative guilty only if such a conclusion can be drawn beyond a shadow of a doubt; the evidence must be strong enough that the null is not true. Statistical hypothesis testing is also firmly based on the idea of rejecting the null hypothesis only if there is strong evidence against it.

For any test of hypotheses, two types of errors are possible, type I errors and type II errors. A *type I error* occurs if the null hypothesis is rejected when it is true. For the trial example, a type I error is committed if the jury declares an innocent person guilty. A *type II error* occurs if the null hypothesis is not rejected, but it is false (the alternative is true). If a guilty person is declared innocent, then a type II error has been made in a jury trial.

In most hypothesis testing settings, we can never be absolutely sure whether the null or the alternative is true. Again thinking of the jury trial, we can never be certain whether she is innocent or guilty. (A confession may not be true; a witness may lie.) The null hypothesis that the person is innocent is rejected only if the evidence presented in the case makes jurors believe that the likelihood of having that much or more evidence against her is extremely unlikely if she is innocent. In *statistical hypothesis testing*, the p-value is the probability of observing an outcome as unusual or more unusual than that that was observed given that the null hypothesis is true. If the p-value becomes too small, then we reject the null hypothesis in favor of the alternative, just as the jurors would reject the hypothesis of innocence and conclude guilty. However, in the case of the statistical test, we have a number, the p-value, which gets smaller as the evidence against the null hypothesis increases.

The significance level of a test, or the α level, is the borderline for deciding whether the p-value is small enough to justify rejecting the null hypothesis in favor of the alternative hypothesis. If the p-value is smaller than the significance level, the null hypothesis is rejected; otherwise, it is not. The significance level is the largest acceptable probability of a type I error.

The standard for rejecting the null hypothesis is quite high. For a jury to declare the defendant guilty, the evidence must be strong enough to remove any shadow of a doubt from the minds of the jurors. As a consequence, some guilty people will be found not guilty only because that shadow of the doubt remained. Similarly, if the p-value is above the significance level and the null is not rejected, this does *not* mean that the null hypothesis is true; we simply do not have enough evidence to say it is not true. This is why, in statistical hypothesis testing, we say that we have "failed to reject the null hypothesis," but we would *not* conclude that the null hypothesis is true.

Example

A farmer wants to determine whether or not his field needs to be treated to control the number of insects in it. Because of costs and environmental concerns, he wants to be sure that treatment is required before proceeding. Based on this, answer the following questions.

1. State the null and alternative hypotheses.
2. How could the farmer make a type I error and what would the consequences be?
3. How could the farmer make a type II error and what would the consequences be?

Solution

1. Because the farmer wants to be certain that treatment is necessary before treating to control insects, this action (of treatment) is the alternative hypothesis. Thus, the null hypothesis is H_0: Do not treat the field, and the alternative hypothesis is H_a: Treat the field.
2. A type I error would occur if the farmer treated the field when it should not have been treated

(rejected a true null hypothesis). This would cause him to spend money unnecessarily on treatment, reducing his profits for the season. The potential for negative environmental impacts is also present.

3. A type II error would occur if the farmer did not treat the field when it should have been treated (failed to reject a false null hypothesis). This would result in lower production and thus a reduction in profits.

▶ Practice

A woman woke up this morning feeling sick to her stomach. She is not sure whether or not she needs to see the doctor. She has no insurance, so she wants to be sure that she really needs medical attention before going to the doctor. Based on this, answer the following questions.

1. State the null and alternative hypotheses.
2. How could the woman make a type I error and what would the consequences be?
3. How could the woman make a type II error and what would the consequences be?

▶ Conducting Hypothesis Tests on Proportions

We will follow five steps in conducting a hypothesis test. Each of these steps will be discussed and applied to proportions in this section.

Step 1: Specifying the Hypotheses

In science, a research hypothesis is a specific, testable prediction made about outcomes of a study. The scientist hopes that the results of the study validate the prediction. When establishing a set (H_0 and H_a) of statistical hypotheses, the research hypothesis is made the alternative hypothesis. To understand why, think back to the parallel we have been using with a jury trial. If the jury believes that strong evidence exists against the assumption of not guilty, then they reject that assumption and conclude that the defendant is guilty. It is not necessary to prove innocence, it is only necessary to raise doubt about guilt to conclude not guilty. Similarly, with statistical hypotheses, if we reject the null hypothesis, we accept the alternative hypothesis. If sufficient evidence does not exist to reject the null hypothesis, we do *not* accept the null; we fail to reject it, which is a much weaker conclusion.

When working with proportions, the null hypothesis is that the population proportion p is equal to some proportion p_0. The alternative may be that p is less than, greater than, or equal to p_0, depending on what the research hypothesis is.

Step 2: Verify Necessary Conditions for a Test and, if Satisfied, Construct the Test Statistic

The conditions for testing hypotheses about the population proportion p are the same as those for constructing a confidence interval on this parameter. They are (1) the sample was randomly selected and (2) the sample is sufficiently large, that is, $np \geq 10$ and $n(1 - p) \geq 10$.

By the Central Limit Theorem, if n is sufficiently large, the sample proportion $\widehat{p}$ is approximately normally distributed with mean p and standard deviation $\sqrt{\dfrac{p(1 - p)}{n}}$. Standardizing $\widehat{p}$, we have $\dfrac{\widehat{p} - p}{\sqrt{\dfrac{p(1 - p)}{n}}}$ is approximately standard normal. If the null hypothesis is true and $p = p_0$, the test statistic

$z_T = \dfrac{\hat{p} - p_0}{\sqrt{\dfrac{p_0(1 - p_0)}{n}}}$ is approximately distributed as

a standard normal random variable. Note: The test statistic is always constructed assuming that the null hypothesis is true.

Notice that the test statistic has the form
$$\frac{point\ estimate - hypothesized\ value}{standard\ error}.$$
The test statistics we will encounter in this book all have this form. They are standardized random variables whose distributions we know *if* the null hypothesis is true.

Step 3: Find the *p*-Value Associated with the Test Statistic

If the null hypothesis is true, the test statistic has an approximate standard normal distribution. If the null hypothesis is not true, the test statistic is not distributed as an approximate standard normal and is more likely to assume a value that is "unusual" for a random observation from a standard normal. The *p*-value is the probability of determining the probability of observing a value as extreme or more extreme as z_T from a random selection of the standard normal distribution.

How do we measure how unusual a test statistic is? It depends on the alternative hypothesis. These are summarized in Figures 16.1, 16.2, and 16.3.

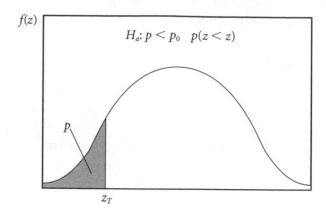

Figure 16.1

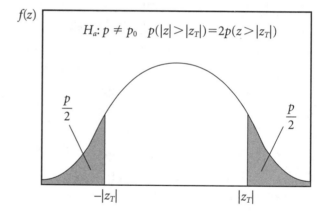

Figure 16.2

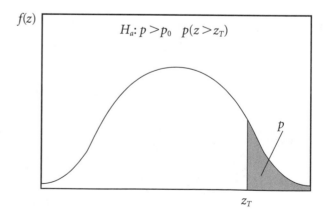

Figure 16.3

Step 4: Decide Whether or Not to Reject the Null Hypothesis

Before beginning the study, the significance level of the test is set. The significance level a is the largest acceptable probability of a type I error. If the p-value is less than α, the null hypothesis is rejected; otherwise, the null is not rejected. In statistical hypothesis testing, we control the probability of a type I error. We often do not know the probability of a type II error. If we reject the null hypothesis, we know the probability of making an error. If we do not reject the null hypothesis, we do not know the probability of having made an error. This is why we would not accept the null hypothesis; we only fail to reject it. In science, a significance level of $\alpha = 0.05$ is generally the standard. That is, we reject H_0 if $p < \alpha = 0.05$. A p-value less than $\alpha = 0.01$ is usually viewed as highly significant. However, the researcher can set the significance level that is most appropriate for his or her study. Once the decision is made to reject or not to reject the null hypothesis, it is important to state what conclusions have been drawn.

Step 5: State Conclusions in the Context of the Study

Statistical tests of hypotheses are conducted to determine whether or not sufficient evidence exists to reject the null hypothesis in favor of the alternative hypothesis. Once the decision is made to reject or not to reject the null hypothesis, it is important to state what conclusions have been drawn.

Example

A sleep researcher believes that most (more than half) of all college students take naps during the afternoon or early evening. He randomly selects 60 students from a large university. He asks each selected student, "Do you regularly take naps during the afternoon or early evening?" Of the 60 students, 34 responded yes. Does sufficient statistical evidence exist to conclude that more than half of the students at this university regularly take afternoon or early evening naps?

Solution

We will follow the five steps of hypothesis testing.

STEP 1: SPECIFYING THE HYPOTHESES

The parameter of interest in the study is p, the proportion of students at this university who regularly take afternoon or early evening naps. The sleep researcher believes more than 50% of the students take afternoon or early evening naps regularly, so this is the alternative hypothesis. Thus, the set of hypotheses to be tested are:

$$H_0: p = 0.50$$
$$H_a: p > 0.50$$

Note: Equality appears in H_0. This is necessary to know the distribution of the test statistic under H_0. Also, a one-sided alternative ($p > 0.50$) is used instead of a two-sided alternative ($p \neq 0.50$). The reason for this is that the sleep scientist wants to conclude that more than half of the students take naps, not that some proportion other than half take naps (the meaning of the two-sided alternative here).

STEP 2: VERIFY NECESSARY CONDITIONS FOR A TEST AND, IF SATISFIED, CONSTRUCT THE TEST STATISTIC

The sleep researcher took a random sample of students from those attending the university, so the first condition for inference is satisfied. To check the second condition, we have $n\widehat{p} = 60\left(\dfrac{34}{60}\right) = 34 > 10$ and $n(1 - \widehat{p}) = 60\left(1 - \dfrac{34}{60}\right) = 26 > 10$. Thus, the second condition for inference is also satisfied. Note that $\widehat{p} = \dfrac{34}{60} = 0.567$.

The test statistic is

$$z_T = \frac{\widehat{p} - p_0}{\sqrt{\dfrac{p_0(1 - p_0)}{n}}}$$

$$= \frac{0.567 - 0.50}{\sqrt{\dfrac{0.50(1 - 0.50)}{60}}} = 1.03.$$

STEP 3: FIND THE *P*-VALUE ASSOCIATED WITH THE TEST STATISTIC

If the null hypothesis is true, the test statistic has an approximate standard normal distribution. The *p*-value is the probability of determining the probability of observing a value as extreme or more extreme as z_T from a random selection of the standard normal distribution. For this study, there would have been more support for the alternative if the sample proportion of nap-taking students had been greater than the observed $\widehat{p} = 0.0567$. This would have led to a larger value of z_T. Thus, the *p*-value is:

$$p = P(z > z_T)$$
$$= P(z > 1.03)$$
$$= 1 - P(z \leq 1.03)$$
$$= 1 - 0.8485$$
$$= 0.1515$$

The 0.8485 was obtained from the standard normal table in Lesson 11 (see Figure 16.4).

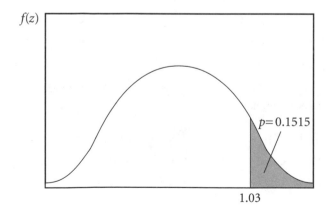

Figure 16.4

STEP 4: DECIDE WHETHER OR NOT TO REJECT THE NULL HYPOTHESIS

If the null hypothesis were true, then we would expect to see a test statistic this extreme or more extreme about 15% of the time. This is not very unusual. Flipping a coin three times and obtaining all heads occurs less frequently (12.5%). If we use any traditional significance level, such as $\alpha = 0.05$ or 0.01, then the *p*-value would be greater than the significance level. For all of these reasons, we would not reject the null hypothesis.

STEP 5: STATE CONCLUSIONS IN THE CONTEXT OF THE STUDY

There is not sufficient evidence to conclude that more than half of the students at this large university regularly take afternoon or evening naps. It is important to note that $\widehat{p} = 0.567$ is greater than 50%, so the sample is consistent with the null hypothesis. However, there is a possibility that $p = 0.50$ and sampling variability caused $\widehat{p}$ to be this large. If $\widehat{p} = 0.567$ is enough larger than 0.50 to be practically important, the researcher may choose to conduct another study using a larger sample size.

▶ Practice

4. A large city is contemplating a ban on smoking in public spaces. The city council wants to institute such a ban only if more than 70% of the adults living in the city support it. To find out if such support exists, the city manager randomly selects 150 of the adult residents in the city and asks them whether or not they support the ban. Of the 150 citizens, 108 support the ban and 42 do not. Is there sufficient statistical evidence to conclude that the strong support for the ban is present among the city's residents? Be sure to follow the five steps of hypothesis testing.

▶ In Short

Statistical hypothesis testing is a fundamental tool in research today. The investigator takes the research hypothesis as the alternative hypothesis if at all possible. Two types of errors are possible. If a true null hypothesis is rejected, a type I error has been committed. If we do not reject a false null hypothesis, a type II error has been committed. The probability of a type I error is controlled. There are five steps to conducting a statistical hypothesis test. It is important to carefully complete each step.

LESSON

17 ▶ Confidence Intervals and Tests of Hypotheses for Means

LESSON SUMMARY

The basic ideas of confidence intervals and hypothesis testing were introduced in Lessons 15 and 16, respectively. These concepts were used to develop large sample confidence intervals and hypothesis tests for population proportions. Here, we will learn how these ideas apply when we are interested in the population mean. Because rarely, if ever, is the population standard deviation known when the population mean is unknown, we will consider only the case where both are unknown.

▶ Confidence Intervals for a Mean

Suppose we have a random sample from a distribution with unknown mean and standard deviation. Further, the distribution may or may not be normal. We would estimate the population mean using the point estimate $\overline{X}$. How would we set the confidence interval on the population mean? First, we need to determine whether the conditions for the statistical methods used here have been met.

To use the statistical methods we will present here to set a confidence interval on the population mean, two conditions must be satisfied for the methods to be valid. (1) The sample must be randomly selected from the population. (2) The population distribution must be normal, or the sample size must be large enough (at least 30) to assume the sampling distribution of $\overline{X}$ is approximately normal by the Central Limit Theorem.

In Lesson 14, we learned that the general form of a confidence interval is *point estimate ± multiplier × standard error*, and this is the form we will use. The point estimate of the population mean is $\overline{X}$, and the standard error of $\overline{X}$ is $\frac{s}{\sqrt{n}}$. Because $t = \dfrac{\overline{X} - \mu}{\frac{s}{\sqrt{n}}}$, t^* with $(n-1)$ degrees of freedom is the multiplier. Thus, the form of the confidence interval for a sample mean is:

$$\overline{X} \pm t^* \frac{s}{\sqrt{n}}$$

if the population distribution is normal with known standard deviation. The choice of t^* depends on the confidence level chosen, just as z^* did when determining the multiplier for proportions. For a $100(1 - \alpha)\%$ confidence interval, t^* is the *t*-value for which $P(t > t^*) = \frac{\alpha}{2}$, where *t* is a random selection from a *t*-distribution with $(n-1)$ degrees of freedom. (By having $\frac{\alpha}{2}$ in each tail, a total of α is in the tails, leaving $1 - \alpha$ for the confidence level. See Figure 17.1.)

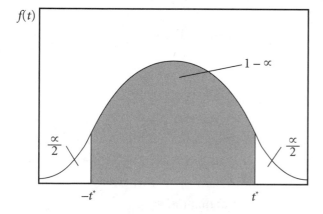

Figure 17.1

Example

A particular species of finch exhibits a polymorphism in bill size that is unrelated to gender. There is a large-billed morph and a small-billed morph. The bill widths within a morph are normally distributed. A random sample of 20 finches is selected from a 100-hectare study region. The average bill width of the sampled finches is 15.92, and the sample standard deviation is 0.066 millimeter. Set a 95% confidence interval on the mean bill width of this species of finch within the study region.

Solution

Because the sample was selected randomly from a normal distribution, the conditions for inference are satisfied. The point estimate of the mean bill width is $\overline{X} = 15.92$. We are given that $s = 0.066$ and $n = 20$, so the standard error of $\overline{X}$ is $\frac{s}{\sqrt{n}} = \frac{0.066}{\sqrt{20}} = 0.0148$.

For a 95% confidence interval, the *t*-value corresponding to a *t* with $n - 1 = 19$ degrees of freedom and $\frac{0.05}{2} = 0.025$ probability in each tail is found by looking in the *t*-table in Lesson 12. The column for 0.025 probability in the upper tail and the column for 19 degrees of freedom intersect to give $t^* = 2.093$. Thus, the confidence interval limits are $15.92 \pm 2.093 \times \frac{0.066}{\sqrt{20}}$ or 15.92 ± 0.03. We are 95% confident that the mean bill length of the finch species is within 0.03 mm of 15.92 mm in the study region.

Example

A study was conducted to determine the mean amount that hair grows daily. A sample of 35 young adults (from 18 to 35 years of age) was randomly selected from all patrons at a large hair salon. The amount of daily hair growth was recorded for each person in the sample. The sample mean was 0.35 mm, and the sample standard deviation was 0.05 mm. Set a 90% confidence interval on the mean daily hair growth of young adults. Interpret the interval in the context of the problem.

Solution

First, we need to determine whether the conditions for inference are satisfied. The first condition that the sample be a random one is satisfied. Although we are not told that the population distribution of daily hair growth is normally distributed, this would seem to be a reasonable assumption. If the data were available, we could construct a histogram and a boxplot to determine whether the assumption of normality is reasonable. However, because $n = 35 > 30$, we can appeal to the Central Limit Theorem and assume that the sampling distribution of $\overline{X}$ is approximately normal.

Because the standard deviation is unknown, we use the following confidence interval:

$$\overline{X} \pm t^* \frac{s}{\sqrt{n}}$$

We are given that $n = 35$, $\overline{X} = 0.35$ mm, and $s = 0.05$ mm. The t^* value we need would have $n - 1 = 34$ degrees of freedom and $\frac{\alpha}{2} = \frac{0.05}{2} = 0.025$ probability in the upper tail. Looking in the t-table in Lesson 12, we have $t^* = 2.032$. The confidence limits are $0.35 \pm 2.032 \frac{0.05}{\sqrt{35}}$ or 0.35 ± 0.02. That is, we are 95% confident that the mean daily hair growth of young adults who are patrons of this salon is between 0.33 and 0.37 mm. Note: Because the population of young adult patrons of this salon was the one from which the sample was selected, it is the one to which we can draw inference. Does the inference extend to all young adults at least in some region about the salon? It does if the patrons of the salon do not differ significantly from others in the region with respect to hair growth. Careful thought should be given before inferences are extended to a broader population than that sampled.

▶ Practice

1. A study was conducted to determine the mean birth weight of a certain breed of kittens. The birth weights of kittens are normally distributed. A sample of 43 kittens was randomly selected from all kittens of this breed born at a large veterinary hospital. The birth weight of each kitten in the sample was recorded. The sample mean was 3.47 ounces, and the sample standard deviation was 0.1 ounces. Set a 95% confidence interval on the mean birth weight of all kittens of this breed. Interpret the interval in the context of the problem.

2. A random sample of 50 lakes was selected from all lakes in a state. The average nitrogen level of these lakes was 754 μg/l, and the sample standard deviation was 18 μg/l. Set a 90% confidence interval on the mean nitrogen level of the lakes in this state.

▶ Hypothesis Testing for a Mean

As with proportions, there are five steps to conducting a test of hypotheses. We will consider each separately.

Step 1: Specifying the Hypotheses

Remember that, when establishing a set (H_0 and H_a) of statistical hypotheses, the research hypothesis is the alternative hypothesis. When working with means, the null hypothesis is that the population mean μ is equal to some value μ_0. The alternative may be that μ is less than, greater than, or equal to μ_0, depending on what the research hypothesis is.

Step 2: Verify Necessary Conditions for a Test and, if Satisfied, Construct the Test Statistic

The conditions for testing hypotheses about the population mean μ are the same as those for constructing a confidence interval on this parameter. If the population distribution is normal, it is enough to know that we have a random sample from the population distribution. If the population distribution is not normal, in addition to having a random sample from the population, the sample size must be large enough to assume that the sampling distribution of $\overline{X}$ is approximately normal (i.e., $n \geq 30$).

If the population is normal and the standard deviation is unknown or if the population distribution is not known but a large sample $(n \geq 30)$ has been selected, the test statistics is $t_T = \dfrac{\overline{X} - \mu_0}{\dfrac{s}{\sqrt{n}}}$ where μ_0 is the value of the mean under the null hypothesis.

Notice that the first test statistics has the form
$$\frac{point\ estimate - hypothesized\ value}{standard\ error}.$$

If the standard deviation of the population and hence the standard deviation of the point estimate are known, we certainly want to use the true value of the parameter instead of the standard error, its estimate. This would lead us to use a different test statistic from that presented here. The reality is that, in practice, the population is rarely, if ever, known, and we will only consider this case.

Step 3: Find the p-Value Associated with the Test Statistic

If the population distribution is normal, the test statistic has a t-distribution with $(n - 1)$ degrees of freedom. Similarly, if the population distribution is not

normal but a large sample has been selected, the test statistic has an approximate t-distribution. If the null hypothesis is not true, the test statistic does not have a t-distribution. The observed value of the test statistic t_T is likely to be unusual for a randomly selected observation from the t-distribution. The p-value continues to be the probability of observing a value as extreme as or more extreme than the test statistic, t_T, from a random selection of an observation from a t-distribution with $(n - 1)$ degrees of freedom.

How do we measure how unusual a test statistic is? It depends on the alternative hypothesis. These are summarized in Table 17.1.

Table 17.1 Alternative hypothesis

ALTERNATIVE HYPOTHESIS	p-VALUE FOR NORMAL POPULATION WITH UNKNOWN VARIANCE OR NONNORMAL POPULATION WITH LARGE SAMPLE
$H_a: \mu < \mu_0$	$P(t < t_T)$
$H_a: \mu = \mu_0$	$P(|t| > |t_T|)$
$H_a: \mu > \mu_0$	$P(t > t_T)$

Step 4: Decide Whether or Not to Reject the Null Hypothesis

Before beginning the study, the significance level α of the test is set. If the p-value is less than the significance level, the null hypothesis is rejected; otherwise, the null is not rejected.

Step 5: State Conclusions in the Context of the Study

Statistical tests of hypotheses are conducted to determine whether or not sufficient evidence exists to reject the null hypothesis in favor of the alternative hypothesis.

Example

A machine is used to produce bolts that have a mean diameter of 1.50 cm. The diameters of the bolts are normally distributed. Sometimes, the machine produces bolts that differ from the desired 1.50 cm mean diameter. If the mean bolt diameter is significantly greater than 1.50 cm, the machine is to be reset. This is expensive because no bolts can be produced while it is being reset. A random sample of 15 bolts is selected from each day's production and tested, and their diameters are recorded. Today, the average diameter of the sampled bolts was 1.52 cm, and the sample standard deviation was 0.01 cm. Is there sufficient evidence to conclude that the machine needs to be reset?

Solution

We follow the five steps of hypothesis testing.

STEP 1: SPECIFYING THE HYPOTHESES

Here, we want strong evidence that the machine needs to be reset. This leads to the following set of hypotheses:

$H_0: \mu = 1.50$

$H_a: \mu > 1.50$

Notice that here we have chosen $\mu = 1.50$ for the null hypothesis because that is the mean diameter of bolts that the machine should be producing. We want to know if the mean diameter becomes significantly greater than 1.50, so $\mu > 1.50$ becomes the alternative hypothesis.

STEP 2: VERIFY NECESSARY CONDITIONS FOR A TEST AND, IF SATISFIED, CONSTRUCT THE TEST STATISTIC

The population distribution of bolt diameters is normal, and the bolts have been randomly selected from the day's production. Thus, the conditions for a test have been satisfied. The test statistic is:

$$t_T = \frac{\overline{X} - \mu_0}{\dfrac{s}{\sqrt{n}}} = \frac{1.52 - 1.50}{\dfrac{0.01}{\sqrt{15}}} = 7.75$$

STEP 3: FIND THE P-VALUE ASSOCIATED WITH THE TEST STATISTIC

Assuming that the population distribution is normal and the standard deviation is known, the test statistic has a t-distribution with $n - 1 = 14$ degrees of freedom *if* the null hypothesis is true. We want to find the probability that a test statistic as extreme or more extreme would be observed if the null hypothesis is not true. Because we are interested only if the sample mean is too large (see the alternative hypothesis), we find

$$p = P(t > t_T)$$
$$= P(t > 7.75) = 1 - P(t \leq 7.75) < 0.001$$

based on the table in Lesson 12.

STEP 4: DECIDE WHETHER OR NOT TO REJECT THE NULL HYPOTHESIS

This p-value is so small that it will be less than any reasonable significance level. Therefore, we would reject the null hypothesis.

STEP 5: STATE CONCLUSIONS IN THE CONTEXT OF THE STUDY

We reject the null hypothesis and conclude that the mean diameter of bolts produced on that day exceeds 1.50 cm. Therefore, the machine needs to be reset.

Example

A researcher has read that U.S. children aged 10 to 17 years watch television an average of 3.6 hours per day. She wants to know if her region differs from this national average. She randomly selects 40 children aged 10 to 17 from the region. Because children do not watch exactly the same amount of television every day, she records the number of hours each sampled child watches television for a four-week period. By taking the average of the hours spent watching television over these 28 days, she has a good measure of the amount of time each child spends watching television daily. The sample mean number of hours these 40 children watched television daily is 2.8 hours, and the sample standard deviation is 2.2 hours. Is there sufficient evidence to conclude that the average number of hours children spend watching television in the researcher's region is different from that reported for the nation?

Solution

Again, we follow the five steps of hypothesis testing.

STEP 1: SPECIFYING THE HYPOTHESES

Here, we want to know whether the average for the region differs from that for the nation. The direction of that difference is not specified, so we have a two-sided alternative; that is, we want to test the following set of hypotheses:

$H_0: \mu = 3.6$

$H_a: \mu \neq 3.6$

STEP 2: VERIFY NECESSARY CONDITIONS FOR A TEST AND, IF SATISFIED, CONSTRUCT THE TEST STATISTIC

A random selection of children was taken from all children in the region (the population of interest). The population distribution of the number of hours children spend watching television daily is very unlikely to be normal. However, because the sample size is 40 ($>$ 30), we know that $\overline{X}$ has at least an approximate normal distribution by the Central Limit Theorem. Therefore, the two conditions for inference are satisfied. The test statistic is

$$t_T = \frac{\overline{X} - \mu_0}{\dfrac{s}{\sqrt{n}}} = \frac{3.6 - 3.2}{\dfrac{2.2}{\sqrt{40}}} = 1.15.$$

STEP 3: FIND THE P-VALUE ASSOCIATED WITH THE TEST STATISTIC

The test statistic has an approximate t-distribution with $n - 1 = 39$ degrees of freedom *if* the null hypothesis is true. We want to find the probability that a test statistic as extreme or more extreme would be observed if the null hypothesis is not true. Because we are interested if the sample mean gets "too far" from the hypothesized population mean, we find $p = P(|t| > |t_T|) = P(|t| > 1.15) > 2 \times 0.1 = 0.2$ using the table in Lesson 12. Notice that the smallest t-value in the line for 39 degrees of freedom in the t-table is 1.304, corresponding to a right-tail probability of 0.01. Because 1.15 is smaller than 1.304, more area is to the right of 1.15 than to the right of 1.304. Thus, $P(t > 1.15) > 0.1$. Because the alternative hypothesis is two sided, this right-tail probability must be doubled to account for the values in the left-tail that would have also been at least as extreme as what we observed if the null hypothesis is true.

STEP 4: DECIDE WHETHER OR NOT TO REJECT THE NULL HYPOTHESIS

The p-value being greater than 0.2 indicates that what we observed would not be at all unusual if the null hypothesis is true ($p > \alpha$). Therefore, we would not reject the null hypothesis.

STEP 5: STATE CONCLUSIONS IN THE CONTEXT OF THE STUDY

We fail to reject the null hypothesis that the mean number of hours children in the researcher's region watch television is the same as that for children in the nation.

▶ Practice

3. Consider again the sample of 50 lakes mentioned earlier in practice problem 2. The state has a large farm economy. A politician has promoted laws that will encourage farmers to adopt practices that will minimize the amount of nitrogen entering the lakes. He has declared that this program has been successful because the mean nitrogen level in the state's lakes does not exceed 750 μg/l. The opposition party would like to say that his claim is not right and that the mean nitrogen level in the state's lakes exceeds 750 μg/l. Recall the average nitrogen level of the 50 sampled lakes was 754 μg/l, and the sample standard deviation was 18 μg/l. Is there evidence that the politician is making a false claim?

▶ In Short

The sample mean is the best estimate of the population mean, but the two are rarely equal. Thus, we use confidence intervals to establish an interval of values that will capture the true population mean with a specified level of confidence. The form of these intervals continues to be *estimate $\pm$ multiplier $\times$ standard error* as it was for proportions. We may also want to test hypotheses concerning values of the population mean. The test statistic for means is $t_T = \dfrac{\overline{X} - \mu_0}{\dfrac{s}{\sqrt{n}}}$, where μ_0 is the hypothesized value of the population mean.

18 ▶ The Matched-Pairs Design for Comparing Two Treatment Means

LESSON SUMMARY

Thus far, we have only attempted to set confidence intervals on proportions or means based on a sample from a single treatment or population. Now we want to conduct studies that will allow us to compare the means of two treatments. First, we will think about how best to design a study. In this lesson, after introducing the basic ideas behind matched pairs and two-group designs, we will focus on the analysis of data from the paired design. In the next lesson, we will consider the two-group design.

▶ Two-Group versus Matched-Pairs Design

Suppose we are going to conduct a study to compare two methods of production, a standard method and a new method, that cause children's dress shoes to shine. Fifty children have been randomly selected to participate in the study. Each child will be given a new pair of dress shoes that shine. But first we need to decide how to assign the treatments (or production methods) to the children's shoes. One approach is to randomly select 25 (half) of the children and give them shoes made using the standard production process; the other half will receive shoes that were made using the new production process. Thus, each child would have a pair of shoes made by one of the two processes. A second approach is to have one shoe of each pair made with the standard process and the other shoe with the new process. Whether the right or left shoe is made with the first process would be randomly determined. In this second approach, each child would wear a dress shoe made using each process.

Regardless of which approach of assigning treatments is used, the children will wear the shoes whenever they wear dress shoes for six months. At the end of the six months, an evaluator who does not know which shoe received which treatment will score the shine quality of each shoe.

Which method of assigning treatments is better? In this case, having each child wear shoes made by both processes is better. Children differ in their activities while wearing dress shoes. Some may wear them only for special occasions, and their shoes will continue to shine no matter what process was used. Other children run and play in their dress shoes. Their shoes are less likely to continue to shine so the process could make a big difference. By having each child wear a shoe made from each process, both processes are subjected to the same environment (level of play). The difference in shine after six months is due more to the differences in the processes and not to differences in the children. This is an example of a *paired experiment*.

The other design in which half of the children wore dress shoes made by the standard process and half by the new process is a *two-group design*. Although this is a reasonable design, it is not the best for this study. The differences we observe in the shine of the shoes after six months are not due only to differences in processes, but also due to differences in children. This would lead to more variation in the estimated mean differences, making it more difficult to determine which, if either, shine process is better.

In the planning stages of a study, it is always important to consider the best way to randomize treatments to the study units. Pairs should be formed if, by pairing, we can eliminate some of the variability in the response that would otherwise be present. In the blinking study presented first in Lesson 4, for each study participant, the number of blinks in a two-minute time period was measured during normal conversation *and* while playing a video game. Those partici-

pants who tended to blink less than average during normal conversation also tended to blink less than average while playing a video game. Similarly, those who tended to blink more than average during normal conversation tended to blink more than average while playing a video game. By recording the difference in the number of blinks under each treatment for each person, we could eliminate the differences among people, allowing us to more accurately measure the differences between treatments, that is, between normal conversation and video playing.

Sometimes, it is not reasonable for both treatments to be applied to the same person. In this case, we may want to pair by some factor that will help explain the variability in the response. For example, suppose we want to compare two treatments for cholesterol. We could pair patients by their initial cholesterol levels. Those with the highest cholesterol level would be in the first pair. Those with the next highest cholesterol level would be in the next pair, and so forth. Then, within each pair, one of the patients would be randomly assigned to the first treatment, and the other would get the second treatment.

Whether or not to use pairing is an important consideration. Matched pairs should be formed only if the researcher believes that significant difference in the response variable can be explained, allowing differences in the treatments to be detected more readily. As an illustration, suppose we decided to pair patients in the cholesterol study on the basis of the length of their feet. The two with the longest feet would be in the first pair, the two with the next longest feet would be in the second pair, and so on. We have no reason to believe that foot length is in any way related to cholesterol level. Pairing in such a situation provides no benefit and is not as effective for assessing whether or not the treatment means are different as the two-group design.

Example

A researcher wants to compare the quality of cooking roasts using two methods—open pan and bag. Four ovens are available for the study. Eight roasts of equal quality have been allocated for the study.

1. Describe how to conduct the study using a matched-pairs design.
2. Describe how to conduct the study using a two-group design.
3. Which of the two designs would you use for this study? Explain.

Solution

1. For a matched-pairs design, two roasts would be cooked in each oven, one in an open pan and the other in a bag. The location of each roast within the oven would be randomly determined.
2. For a two-group design, we randomly select two of the ovens to cook a roast using the open-pan method; the other two ovens would each be used to cook a roast in a bag.
3. The matched-pairs design would be the best for this study. Ovens often vary in their ability to hold temperature at a specified level. By having both treatments in each oven, differences between ovens can be accounted for in the analysis. As described, we have used half as many roasts in the two-group design. We could put two roasts in each oven and cook using the same method. This gives us information on the differences within an oven and allows us to more precisely estimate the quality of the roasts cooked in a specific oven. Cooking two roasts in the same oven does *not* double the number of experimental units in the study. An oven would be the experimental unit because the cooking methods were randomly assigned to the ovens.

▶ Practice

A dermatologist wants to know which of two creams is better for curing hand eczema. For the experiment, she will apply the hand creams and determine which one works better. Forty patients with hand eczema volunteer for the experiment.

1. Describe how to conduct the study using a matched-pairs design.
2. Describe how to conduct the study using a two-group design.
3. Which of the two designs would you use for the study? Explain.

▶ Matched-Pairs Design

Once we have decided to conduct an experiment using matched pairs, how do we actually go about conducting the study? First, the study units need to be obtained. As we learned in Lesson 2, if the study units are randomly selected from some population, conclusions can be made for that population at the end of the study; otherwise, conclusions apply only to the units in the study. In the shoe-shine study, children were randomly selected. The group from which these children were randomly selected is the population for which inferences can be made.

Next, the study units need to be paired. Individuals could be matched according to a characteristic that could explain some of the difference in the response variable. In the cholesterol study, individuals were matched by initial cholesterol level. Sometimes, both treatments can be sequentially applied to the same individual. This form of matched pairs is often very strong, but may require more time than is available for the study.

Once the pairs are formed, one treatment is randomly assigned to one unit in the pair; the other unit receives the second treatment. Notice that a separate randomization is used for each pair. For the shoe-shine study, it would not be sufficient to flip a coin and randomly assign the first treatment to all right shoes and the other treatment to all left shoes. Children are right- or left-footed just as they are right- or left-handed. It is possible that one shoe, say the right shoe, tends to get the most wear because most children are right-footed. If this is the case, then the treatment assigned to all right shoes would be at a disadvantage in the study. To avoid this and other biases of which we may not even be aware, we randomly assign treatments within each pair.

Once the study is complete, we record the response variable for each unit. Let X_{1i} be the observed response from the first treatment in pair $i, i = 1, 2, \ldots,$ n, where there are n pairs. Similarly, let X_{2i} be the observed response from the second treatment in pair $i, i = 1, 2, \ldots, n$. Then $D_i = X_{1i} - X_{2i}, i = 1, 2, \ldots,$ n, is the observed difference in the two treatments for the ith pair. There is a conceptual population of D_i's comprised of the differences in all possible pairs that could have been used in this study. This population has μ_D and standard deviation σ_D.

The sample mean difference in the two treatments, $\overline{D} = \dfrac{1}{n}\sum_{i=1}^{n} D_i = \overline{X}_1 - \overline{X}_2$, is an estimate of the difference in the treatment means, $\mu_1 - \mu_2 = \mu_D$, the mean of the population of paired treatment differences. The sample variance of the pairwise differences provides an estimate of σ_D^2 and is $s_D^2 = \dfrac{1}{n-1}\sum_{i=1}^{n} (D_i - \overline{D})^2$. The sample standard deviation is $s_D = \sqrt{s_D^2}$. Notice that $\overline{D}$ and s_D are, respectively, the sample mean and sample standard deviation of

the differences. This would lead us to speculate that the standard error of $\overline{D}$ is $\dfrac{s_D}{\sqrt{n}}$. This is, in fact, the case! The analysis of a paired study is based on these quantities. We will consider this further in the next two lessons.

Example

An athletic shoe company believes that they have developed a shoe that will help short-distance runners lower their times in races. They recruited 24 runners. Each runner was given a new pair of the athletic shoes. The runners were encouraged to use these shoes and their favorite pair of running shoes equally in practice for two weeks. After two weeks, the runners ran two 100-meter dashes with five hours between races. For each runner, a coin was flipped. If the coin landed heads up, the runner wore his or her favorite running shoes in the first race; otherwise, he or she wore his or her newly developed shoes. In the second race, each runner wore the pair of shoes that was not used in the first race. The times for the runners are given in Table 18.1.

1. Explain why this study has a matched-pairs design. Include a clear statement describing what constitutes a pair.
2. Find the difference in observations from each pair.
3. Estimate the mean and standard deviation of the differences in time to run a 100-meter dash when wearing the favorite running shoes compared to the new running shoes.
4. Find the standard error of the estimated mean of the differences in time to run a 100-meter dash when wearing the favorite running shoes compared to the new running shoes.
5. Is the assumption reasonable that the differences are normally distributed?
6. To which population may inference be drawn from this study?

Table 18.1 Race times with different shoes

RUNNER	1	2	3	4	5	6	7	8
Favorite Shoes	10.83	10.83	11.23	11.06	10.43	11.39	10.92	11.41
New Shoes	10.89	10.19	11.13	10.75	10.73	11.01	10.74	11.08
RUNNER	9	10	11	12	13	14	15	16
Favorite Shoes	11.05	11.64	10.65	10.40	10.77	10.73	10.77	11.15
New Shoes	10.85	11.17	10.39	10.64	10.81	10.03	10.57	10.70
RUNNER	17	18	19	20	21	22	23	24
Favorite Shoes	10.77	10.91	10.89	10.30	11.08	11.86	10.48	11.00
New Shoes	10.65	11.02	10.84	10.19	10.70	11.02	10.87	10.66

Solution

1. The two treatments are the favorite running shoes and the newly developed running shoes. Each treatment is applied to a runner. Thus, the favorite running shoes and the newly developed running shoes are paired by runner. A pair consists of the running times for the two treatments from a single runner. The order in which the shoes were used was randomized for each runner, a critical step in conducting the study.

2. The differences in the two treatments are computed for each runner (see Table 18.2).

3. The estimated mean difference in the running times using the favorite shoes versus using the newly developed shoes is

$$\overline{D} = \frac{1}{n}\sum_{i=1}^{n} D_i$$
$$= \frac{1}{24}(-0.06 + 0.64 + \ldots + 0.34)$$
$$= 0.2050.$$

The estimated variance of these differences is $s_D^2 = \frac{1}{n-1}\sum_{i=1}^{n}(D_i - \overline{D})^2 = 0.0941$, and the estimated standard deviation is 0.3068.

4. The standard error of the estimated differences in the two treatments is
$$s_{\overline{D}} = \frac{s_D}{\sqrt{n}} = \frac{0.3068}{\sqrt{24}} = 0.0626.$$

Table 18.2 Race time differences

RUNNER	1	2	3	4	5	6	7	8
Difference	−0.06	0.64	0.10	0.31	−0.30	0.38	0.18	0.33
RUNNER	9	10	11	12	13	14	15	16
Difference	0.20	0.47	0.26	−0.24	−0.04	0.70	0.20	0.45
RUNNER	17	18	19	20	21	22	23	24
Difference	0.12	−0.11	0.05	0.11	0.38	0.84	−0.39	0.34

5. Because the sample size is small, it is difficult to determine whether or not the observed differences are normally distributed. Although formal tests exist for determining normality, we will not study them here. Instead, we will rely on examining graphs to determine whether there are indications that the data may not be normal. Figures 18.1, 18.2, and 18.3 show a histogram, a dotplot, and a boxplot, respectively. The histogram looks fairly symmetric and unimodal. With only 24 observations, the shape of a population is often not fully captured in a histogram of the data. The dotplot appears to be centered at about 0.20. The values range from -0.39 to 0.84 with a higher concentration of dots in the center. From the boxplot, the data appear to be fairly symmetric without any outliers. In summary, we do not see any indication of skewness, outliers, or other features that would cause us to think that the assumption of normality is unreasonable.

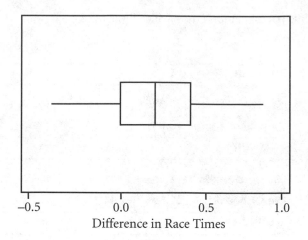

Figure 18.3

6. Because the runners were recruited and not randomly selected from some population, the population to which inference may be drawn is the runners in this study.

▶ Practice

A high school chemistry teacher has given his students an in-class assignment. He has given each of his students two liquids, *A* and *B*, and told the students that one of the liquids is saltwater and the other is plain water, but not which one is which. The students have learned that, when adding salt to water, the boiling point increases. The teacher has asked his students to record each liquid's boiling point and, based on this information, determine which liquid is saltwater and which is plain water. Fifteen students are in the class, and each student has a burner with two side-by-side hot plates on it. Each student flips a coin. If the coin lands heads up, the student places liquid *A* on the right hot plate. If the coin lands tails up, the student places liquid *B* on the right hot plate. The students use glass thermometers to record the temperatures of the liquids at the exact time they begin to boil. The temperatures were recorded in degrees Celsius and are given in Table 18.3.

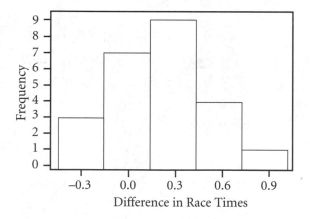

Figure 18.1

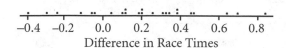

Figure 18.2

Table 18.3 Record of boiling points

STUDENT	1	2	3	4	5
Liquid A	109.1	107.0	106.3	104.9	106.8
Liquid B	100.6	100.2	99.8	98.3	99.3
STUDENT	6	7	8	9	10
Liquid A	110.9	108.3	105.7	109.8	108.9
Liquid B	102.1	101.4	99.9	101.1	100.9
STUDENT	11	12	13	14	15
Liquid A	107.9	104.9	108.8	109.3	105.8
Liquid B	100.8	98.5	102.0	100.9	100.3

4. Explain why this study has a matched-pairs design. Include a clear statement describing what constitutes a pair.
5. Find the difference in observations from each pair.
6. Estimate the mean and standard deviation of the differences in the boiling points of each liquid.
7. Find the standard error of the estimated mean of the differences in the boiling points of each liquid.
8. Is the assumption that the differences are normally distributed reasonable?
9. To which population may inferences be drawn from this study?

Confidence Intervals on the Difference in Two Treatment Means

If the goal is to provide an interval of reasonable values for the mean difference in two treatment means based on a matched-pairs design, we want to set a confidence interval on that mean difference. The conditions for inference are that the differences (D_i's) are a random sample from a population of differences. Second, the D_i's are either normally distributed or the sample size is large enough (at least 30) to assume that the average of the D_i's is approximately normally distributed by the Central Limit Theorem.

The methods for statistical inference using the D_i's are computationally identical to those for one population; the interpretation is all that differs. Recall $\overline{D} = \dfrac{1}{n}\sum_{i=1}^{n} D_i = \overline{X}_1 - \overline{X}_2$ is an unbiased estimate of the difference in the treatment means, $\mu_1 - \mu_2 = \mu_D$. The standard error of this estimate is $s_{\overline{D}} = \dfrac{s_D}{\sqrt{n}}$ where $s_D^2 = \dfrac{1}{n-1}\sum_{i=1}^{n}(D_i - \overline{D})^2$. Using the form *point estimate $\pm$ multiplier $\times$ standard error*, a $100(1 - \alpha)\%$ confidence interval on $\mu_D = \mu_1 - \mu_2$ has the form $\overline{D} \pm t^\star \dfrac{s_D}{\sqrt{n}}$ where $t^\star$ with $(n - 1)$ degrees of freedom is the proper tabulated value to give $100(1 - \alpha)\%$ confidence.

Example

Look again at the study comparing running shoes earlier in this lesson. Find a 90% confidence interval on the mean difference in the race times for a 100-meter dash using the athlete's favorite running shoes and the new running shoes.

Solution

The 24 runners were recruited not randomly selected from all runners, so our inference will be restricted to these 24 runners (review the table in Lesson 2). We may believe that these runners are representative of all runners and thus attempt to broaden our scope of inference, but we need to be very careful in doing so. If we think these runners may differ from a broader population of runners, the 24 runners must be taken as the population of interest. The random assignment of treatment order is necessary for the first condition of inference to be satisfied. Note: Unless treatments are randomly assigned within a pair, we do not have a random sample of differences, which is the first condition. Although we have not formally tested whether or not the population of differences has a normal distribution, the graphs constructed in the previous example suggest that it is not an unreasonable assumption, so we will assume that the differences in running times are at least approximately normally distributed.

Based on $n = 24$ runners, we found $\overline{D} = 0.2050$ and $s_D = 0.3068$. For a 90% confidence interval, $\alpha = 0.10$, so we want to put 5% of the probability in each tail. Looking in the t-table in Lesson 12, the t-value at the intersection of the row corresponding to 23 degrees of freedom and the column showing 0.05 in the upper tail, we have $t^\star = 1.714$. The confidence interval on $\mu_F - \mu_N$ is $0.2050 \pm 1.714\dfrac{0.3068}{\sqrt{23}}$ or 0.2050 ± 0.1096. Therefore, we estimate that, on average, the new running shoes allow runners to complete the race in 0.21 fewer seconds compared to their favorite running shoes, and we are 90% confident that this estimate is within 0.11 seconds of the true mean difference in the times to run the 100-meter dash using the new running shoes and the runners' favorite shoes.

▶ Practice

10. Look again at the study of the temperatures at which liquids A and B boiled in the previous practice problem. Find a 95% confidence interval for the mean difference in the temperatures at which liquid A and liquid B boiled. Be sure to interpret the interval in the context of the problem.

▶ Hypothesis Tests Concerning the Difference in Two Treatment Means

Tests of hypotheses concerning the difference in two treatment means are based on the same philosophy as the hypothesis tests discussed earlier in this book. Five steps are followed to conduct a hypothesis test.

Step 1: Specifying the Hypotheses

For a matched-pairs design, the null hypothesis, H_0, is that the difference in the treatment means is d_0; that is, $H_0: \mu_1 - \mu_2 = d_0$. Although it is common for $d_0 = 0$ (implying that the means or proportions are equal), this is not necessary; d_0 can be any value. The alternative is that the difference is less than, greater than, or equal to d_0.

Step 2: Verify Necessary Conditions for a Test and, if Satisfied, Construct the Test Statistic

The conditions for testing hypotheses about the difference in two treatment means are the same as conditions for testing confidence intervals. First, the differences (D_i's) must be a random sample from some population of differences. Second, we must satisfy the condition of normality. That is, the D_i's are normally distributed or the sample size is sufficient that the sample mean difference is approximately normal by the Central Limit Theorem.

The test statistic has the now familiar form $$\frac{point\ estimate\ -\ hypothesized\ value}{standard\ error}.$$

For the paired design, this becomes

$$t_T = \frac{\overline{D} - d_0}{\dfrac{s_D}{\sqrt{n}}}.$$

We know the distribution of this test statistic is at least approximately a t-distribution with $(n - 1)$ degrees of freedom *if* the null hypothesis is true.

Step 3: Find the *p*-Value Associated with the Test Statistic

If the null hypothesis about the difference in two treatment means is true, the test statistic has either an exact or an approximate t-distribution. The distribution of the test statistic when the null hypothesis is true is called the *null distribution*. If the null hypothesis is not true, the test statistic is not distributed according to the null distribution and is more likely to assume a value that is "unusual" for a random observation from that distribution. The p-value is the probability of determining the probability of observing a value as extreme as or more extreme than the test statistic from a random selection of the standard normal distribution.

How do we measure how unusual a test statistic is? It depends on the alternative hypothesis. These are summarized in Table 18.4.

Table 18.4

ALTERNATIVE HYPOTHESIS	P-VALUE
$H_a: \mu < \mu_0$	$P(t < t_T)$
$H_a: \mu = \mu_0$	$P(|t| > |t_T|) = 2P(t > |t_T|)$
$H_a: \mu > \mu_0$	$P(t > t_T)$

Step 4: Decide Whether or Not to Reject the Null Hypothesis

Before beginning the study, the significance level of the test is set. The significance level is the largest acceptable probability of a type I error. If the p-value is less than the significance level, the null hypothesis is rejected; otherwise, the null is not rejected.

Step 5: State Conclusions in the Context of the Study

Statistical tests of hypotheses are conducted to determine whether or not sufficient evidence exists to reject the null hypothesis in favor of the alternative hypothesis.

Example

For the running shoe study presented in the previous lesson, the company wants to be able to claim that the new shoe reduces the mean of the 100-meter race times for runners. Is there statistical evidence to support this claim?

Solution

STEP 1: SPECIFYING THE HYPOTHESES

Let the subscripts F and N represent the runners' favorite shoes and the newly developed shoes, respectively. The company wants to know whether the mean of the race times is lower for the new shoes. (Learning that the mean is greater would certainly not be a strong promotional point.) Thus, the hypotheses of interest are H_0: $\mu_F - \mu_N = 0$ versus H_0: $\mu_F - \mu_N > 0$. Notice we could have written these as H_0: $\mu_F = \mu_N$ and H_0: $\mu_F > \mu_N$. The two sets of hypotheses are equivalent.

STEP 2: VERIFY NECESSARY CONDITIONS FOR A TEST AND, IF SATISFIED, CONSTRUCT THE TEST STATISTIC

Although the runners were recruited, the order in which the treatments (newly developed shoes and favorite shoes) were observed was randomly determined. Thus, the observed differences are a random sample of all possible differences for these 24 runners, and the first condition for inference is satisfied. Often, race times are normally distributed, so the differences in race times under two treatments would be normally distributed. The sample size is not sufficient to test the assumption of normality rigorously. However, from inspection of the graphs and summary statistics earlier in this lesson, it is not unreasonable to assume that the differences in run times using the favorite and new shoes are normally distributed. Thus, the conditions for inference are assumed to be satisfied.

Because the study has a paired design, the test statistic is the now familiar form

$$t_T = \frac{\overline{D} - d_0}{\dfrac{s_D}{\sqrt{n}}} = \frac{0.2050 - 0}{\dfrac{0.3068}{\sqrt{24}}} = 3.27.$$

STEP 3: FIND THE *P*-VALUE ASSOCIATED WITH THE TEST STATISTIC

If the null hypothesis is true, the test statistic has a t-distribution with $(n - 1) = 23$ degrees of freedom. Given the alternative hypothesis, we want to reject the null hypothesis if t_T gets too large. Thus, $p = P(t > 3.27)$. In the t-table on the line for 23 degrees of freedom, 3.27 lies between 2.807 and 3.485, corresponding to upper tail probabilities of 0.005 and 0.001, respectively; thus, $0.001 < p < 0.005$.

STEP 4: DECIDE WHETHER OR NOT TO REJECT THE NULL HYPOTHESIS

The p-value observed in this study indicates that a test statistic of this magnitude is very unusual if the null hypothesis is true. Therefore, we reject the null hypothesis and decide in favor of the alternative.

STEP 5: STATE CONCLUSIONS IN THE CONTEXT OF THE STUDY

The mean time for a 100-meter race was significantly less when runners wore the newly developed shoes compared to their favorite running shoes.

▶ Practice

11. Once again, consider the study of the differences in the boiling point of saltwater and plain water. Based on the student's data, is the mean temperature at which liquid *A* boils significantly different from the mean temperature at which liquid *B* boils? Recall that the boiling point of saltwater is greater than that of plain water. Use this information to answer the science teacher's question, "Which of these two liquids is saltwater?"

▶ In Short

Designs comparing two treatments or populations have been discussed. The matched-pairs design allows one to account for known or suspected sources of variability in the design. The two-group design is useful when a reasonable basis for pairing is not available or feasible. For the paired design, confidence intervals and hypothesis tests on the difference in the treatment means were described.

LESSON 19 ▶ Confidence Intervals for Comparing Two Treatment or Population Means

LESSON SUMMARY

Matched-pairs and two-group designs were considered in the previous lesson, but only the paired design was discussed in detail. Now we will focus on the two-group design and on random samples from two populations. Design considerations as well as inference for the difference in the treatment or population means will be discussed.

▶ Two-Group Designs

Suppose we have decided to conduct a study using a two-group design. As with the paired design, we begin by selecting the study units. If this selection is made at random from some population, inferences can be made for this population at the end of the study. Otherwise, inference will be restricted to units in the study.

After the study units have been chosen, half are randomly assigned to the first treatment; the other half receive the second treatment. It is not necessary for the two groups to be evenly divided as just described. We could flip a fair coin to determine which treatment each unit receives. Although about half would get each treatment, it is likely that one treatment will have a few more study units than the other treatment. There are times in which we want to have more units receiving one treatment than another. However, in the absence of additional information, we will seek a randomization process that will result in the same number of units within each group.

The goal of a two-group study is usually to compare the means of the two groups. Let the mean of the first population be denoted by μ_1 and the mean of the second population by μ_2. Let X_{1i} be the observed ith response from the first treatment, $i = 1, 2, \ldots, n_1$, where n_1 units are receiving treatment 1. Similarly, let X_{2i} be the observed response from the ith unit receiving the second treatment, $i = 1, 2, \ldots, n_2$. We use the sample mean of the units receiving the first treatment to estimate that treatment mean, and the sample mean of the units under the second treatment to estimate the second treatment mean. Let $\overline{X}_1$ and $\overline{X}_2$ be the sample means based on units receiving treatments 1 and 2, respectively.

To estimate the difference in the two treatment means, $\mu_1 - \mu_2$, we would use $\overline{X}_1 - \overline{X}_2$. Although we have only one sample from each treatment, we can imagine repeating the study many times, and computing $\overline{X}_1 - \overline{X}_2$ each time. This gives rise to the sampling distribution of $\overline{X}_1 - \overline{X}_2$. If each population distribution is normal, the sampling distribution of $\overline{X}_1 - \overline{X}_2$ is normal with mean $\mu_1 - \mu_2$ and variance $\dfrac{\sigma_1^2}{n_1} + \dfrac{\sigma_2^2}{n_2}$.

The standard error of $\overline{X}_1 - \overline{X}_2$ depends on whether the variances of the units receiving the two treatments are equal ($\sigma_1^2 = \sigma_2^2$). If we believe that the two variances are equal, then we want to use information from each sample to estimate the common variance; that is, we want to find the pooled estimate of the variance. The term *pooled estimate* means that information from multiple samples is combined to provide one estimate. We must allow for the fact that the means could be different under the two treatments. These ideas lead us to use s_p^2 (called s-squared pooled),

$$s_p^2 = \frac{(n_1 - 1)s_1^2 + (n_2 - 1)s_2^2}{n_1 + n_2 - 2}$$

as the estimate of this common variance. Notice that s_p^2 is a weighted average of the estimated variances within

each treatment. If the two samples sizes (n_1 and n_2) are equal, s_p^2 is the average of s_1^2 and s_2^2; otherwise, the sample having the largest number of observations has the largest weight. Assuming that the variance is the same under the two treatments, the standard error of $\overline{X}_1 - \overline{X}_2$ is $\sqrt{\dfrac{s_p^2}{n_1} + \dfrac{s_p^2}{n_2}} = s_p\sqrt{\dfrac{1}{n_1} + \dfrac{1}{n_2}}$.

What happens if we are unwilling to assume that the variances are the same under the two treatments? In this case, we must obtain estimates of the variance for units receiving each treatment. That is, s_1^2 is the sample variance for all units receiving treatment 1. Similarly, s_2^2 is the sample variance for all units receiving treatment 2. The standard error of $\overline{X}_1 - \overline{X}_2$ is

$$\sqrt{\frac{s_1^2}{n_1} + \frac{s_2^2}{n_2}}.$$

Example

A large telemarketing firm acquired a new client with a product. A script for the sales people to use when calling prospective customers needed to be developed. Because the product was different from ones the firm had handled in the past, the script writers were divided as to which of two approaches, a hard-sell approach or a soft-sell approach, would result in the greatest number of sales. They decided to conduct a study to compare the two approaches. Eighty people were randomly selected from the sales force. Of these, 40 were randomly assigned to use the hard-sell approach; the other 40 were to use the soft-sell approach. Each person was then trained using the script of the method to which he or she was assigned. After having each study participant use the script for one day, the number of sales made during a randomly selected hour during the next work day was recorded. The results are in Table 19.1.

1. This study has a two-group design. Explain why this statement is true.

2. Estimate the mean and standard deviation for each treatment.

3. Is it reasonable to assume the variance is the same for both populations? If so, estimate the variance common to both.

4. Estimate the difference in the treatment means and find its standard error.

5. Is it reasonable to assume that the numbers of sales are normally distributed for each treatment?

6. To which population may inference be drawn from this study?

Solution

1. The hard-sell approach was randomly assigned to half of the study participants, and the other half of the study participants was assigned the soft-sell approach. No effort was made to pair the study participants to control other factors.

2. The estimated mean number of sales per hour when using the hard-sell approach is 1.45 sales. The estimated variance of the number of sales per hour for this approach is 1.59 sales^2, so the estimated standard deviation is 1.26 sales. The estimated mean and variances of the number of sales per hour when using the soft-sell approach is 2.38 sales and 1.93 sales^2, respectively. The estimated standard deviation of the number of sales using the soft-sell approach is 1.39 sales.

3. Because the standard deviations for the two treatment groups are similar, it is reasonable to assume that they are estimating a common variance.

Table 19.1

PARTICIPANT	1	2	3	4	5	6	7	8
Hard-Sell Sales	2	0	1	2	3	1	2	1
Soft-Sell Sales	4	4	4	1	2	1	1	3
PARTICIPANT	**9**	**10**	**11**	**12**	**13**	**14**	**15**	**16**
Hard-Sell Sales	1	0	1	2	1	0	0	0
Soft-Sell Sales	4	2	3	3	2	2	0	3
PARTICIPANT	**17**	**18**	**19**	**20**	**21**	**22**	**23**	**24**
Hard-Sell Sales	2	0	0	0	1	3	4	0
Soft-Sell Sales	3	4	4	5	1	0	3	2
PARTICIPANT	**25**	**26**	**27**	**28**	**29**	**30**	**31**	**32**
Hard-Sell Sales	1	3	1	3	1	2	1	2
Soft-Sell Sales	4	3	4	3	0	2	1	1
PARTICIPANT	**33**	**34**	**35**	**36**	**37**	**38**	**39**	**40**
Hard-Sell Sales	2	1	0	5	2	4	2	1
Soft-Sell Sales	1	1	2	3	4	0	1	3

Using the subscript H to represent the hard-sell approach and the subscript S to represent the soft-sell approach, the estimate of that common variance is:

$$s_P^2 = \frac{(n_H - 1)s_H^2 + (n_S - 1)s_S^2}{n_H + n_S - 2}$$

$$= \frac{(40 - 1)(1.59) + (40 - 1)(1.93)}{40 + 40 - 2}$$

$$= 1.76 \text{ sales}^2$$

4. The estimated difference in the mean number of sales using the hard-sell and the soft-sell approaches is $\overline{X}_H - \overline{X}_S = 1.45 - 2.38 = -0.93$ sales; that is, 0.93 fewer sales are made, on average, using the hard-sell approach compared to the soft-sell approach. The standard error of this estimate is:

$$\sqrt{s_P^2\left(\frac{1}{n_N} + \frac{1}{n_S}\right)} = \sqrt{1.76\left(\frac{1}{40} + \frac{1}{40}\right)}$$

$$= 0.30 \text{ sales}$$

5. For a two-group experiment, the condition of normality is checked within each treatment group. Because we are working with counts (the number of sales in an hour), the data are discrete. They cannot be normally distributed. We will focus on the shape of the sample distributions. Figures 19.1 and 19.2 show parallel dotplots and parallel boxplots. Based on the dotplot, the sample distribution of the hard-sell approach is skewed to the right, but the distribution of the soft-sale approach is reasonably symmetric. The boxplot supports the view that the sample distribution of the hard-sell approach is skewed to the right; further, the lone observation of five sales in an hour is an outlier. Based on the boxplot, the symmetry of the sample distribution of the soft-sell approach is a reasonable assumption though some may believe the distribution to be skewed left.

6. Because the study participants were randomly drawn from the firm's sales force, the sales force of this large telemarketing firm is the population to which inference may be drawn.

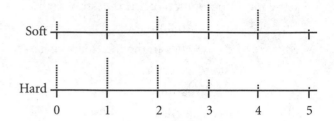

Figure 19.1

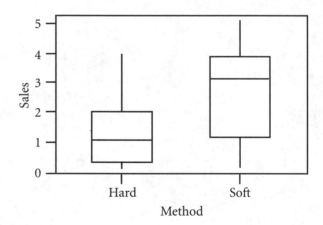

Figure 19.2

▶ Practice

A teacher became curious as to whether or not caffeine had an affect on a person's memory. He began to consider this when a former student of his commented that drinking caffeinated beverages while studying for a test helped him remember the facts for the test. He decided to do a study and randomly selected 100 of his students to participate. He then randomly assigned 50 students to ingest one glass of fruit juice with 45 mg of caffeine and 50 students to drink one glass of fruit juice with no caffeine added. The students did not know whether they drank the juice with caffeine or the juice without it. Thirty minutes after the students drank the juice, the professor had them perform a memory test. Out of a possible 15 points, the students performed as shown in Table 19.2.

1. This study has a two-group design. Explain why this statement is true.
2. Estimate the mean and standard deviation for each treatment.
3. Is it reasonable to assume the variance is the same for both populations? If so, estimate the variance common to both.
4. Estimate the difference in the treatment means and find its standard error.
5. Is the assumption that the memory scores are normally distributed reasonable for each treatment?
6. To which population may inference be drawn from this study?

Table 19.2 Performance with and without caffeine

STUDENT	1	2	3	4	5	6	7	8	9	10
Caffeine	9	9	10	12	13	10	12	13	10	15
No Caffeine	12	13	12	11	10	10	14	13	8	10
STUDENT	11	12	13	14	15	16	17	18	19	20
Caffeine	15	12	11	13	10	13	13	11	13	9
No Caffeine	12	10	13	15	15	12	12	11	11	11
STUDENT	21	22	23	24	25	26	27	28	29	30
Caffeine	14	13	9	11	9	12	11	12	10	13
No Caffeine	9	14	12	14	9	12	12	12	12	11
STUDENT	31	32	33	34	35	36	37	38	39	40
Caffeine	13	11	13	11	15	15	12	11	9	12
No Caffeine	10	7	9	14	23	15	9	9	10	13
STUDENT	41	42	43	44	45	46	47	48	49	50
Caffeine	9	11	13	14	11	11	15	15	10	10
No Caffeine	9	10	13	10	10	11	11	14	10	12

► Comparing Two Populations

If we randomly select samples from each of two populations, the two samples are independent. The statistical methods used to compare the means of two populations based on independent samples from each population are identical to those used in analyzing studies of the two-group design. We estimate the difference using $\overline{X}_1 - \overline{X}_2$. The standard error of $\overline{X}_1 - \overline{X}_2$ is $\sqrt{\dfrac{s_p^2}{n_1} + \dfrac{s_p^2}{n_2}} = s_p\sqrt{\dfrac{1}{n_1} + \dfrac{1}{n_2}}$ if the two population standard deviations are equal. If the two population standard deviations are not equal, then the standard error of $\overline{X}_1 - \overline{X}_2$ is $\sqrt{\dfrac{s_1^2}{n_1} + \dfrac{s_2^2}{n_2}}$.

Example

A researcher conducted a study to compare traits of identical and fraternal twins. She wanted to know whether the mean difference in twin heights was different for identical and fraternal twins. She recruited 30 identical twin pairs and 30 fraternal twin pairs to participate in the study. The difference in each pair's height was recorded and presented in Table 19.3.

1. This study compares two population means. Explain why this statement is true.
2. Estimate the mean and standard deviation for each population.
3. Is it reasonable to assume the variance is the same for both populations? If so, estimate the variance common to both?
4. Estimate the difference in the population means and find the standard error of this estimate.
5. Is it reasonable to assume that the differences in twin heights are normally distributed for each population?
6. To which populations may inference be drawn from this study?

Solution

1. The populations of interest are the population of identical twins and the population of fraternal twins. The type of twins cannot be assigned at random. Fraternal and identical twins constitute different populations.
2. The estimated mean difference in the heights of identical twins is 1.68 cm. The estimated variance of the difference in the heights of identical twins is 2.10 cm², and the estimated standard

Table 19.3 Difference in the height of twins

TWIN	1	2	3	4	5	6	7	8	9	10
Identical	6.3	4.7	2.1	2.8	3.9	3.5	2.4	2.3	0.1	0.3
Fraternal	14.1	10.8	12.9	9.4	8.6	6.6	7.1	7.7	4.0	5.4
TWIN	11	12	13	14	15	16	17	18	19	10
Identical	0.2	1.8	1.6	1.5	1.3	1.6	1.5	1.0	0.5	0.4
Fraternal	4.6	4.9	5.5	2.4	2.7	3.2	3.6	3.8	0.1	0.0
TWIN	21	22	23	24	25	26	27	28	29	30
Identical	0.6	0.8	1.1	0.5	0.9	0.6	1.3	0.1	1.8	2.9
Fraternal	0.6	1.2	1.8	0.9	1.5	0.4	0.3	1.2	1.6	2.8

deviation is 1.45 cm. The estimated mean difference in the heights of fraternal twins is 3.40 cm. The estimated variance of the difference in heights of fraternal twins is 14.76 cm², and the estimated standard deviation is 3.842 cm.

3. The variance of the difference in heights of fraternal twins is about seven times the variance of the difference in heights of identical twins. Thus, it is unlikely that these are estimates of the same quantity. (In general, if one variance is about four times that of the other, then it is unlikely the two are equal.) Thus, we would not want to estimate a common variance.

4. The estimated difference in population means is $\overline{X}_F - \overline{X}_I = 3.9 - 1.5 = 2.4$ cm. Because the variances are not the same, the standard error of the estimate is

$$\sqrt{\frac{s_I^2}{n_1} + \frac{s_F^2}{n_2}} = \sqrt{\frac{2.10}{30} + \frac{14.76}{30}} = 0.75 \text{ cm.}$$

5. Parallel dotplots and boxplots are shown in Figures 19.3 and 19.4. Both graphs indicate that the sample distributions are skewed right. The difference in identical twin heights has an outlier as well. Normality is not a reasonable assumption for these populations.

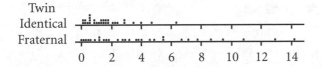

Figure 19.3

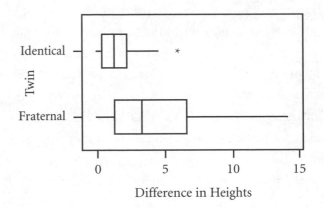

Figure 19.4

6. Because twins were recruited and not randomly selected, inference may be drawn only to twins in the study. We would hope that this sample is representative of the larger population of identical and fraternal twins so that the inferences could be drawn more broadly. However, we cannot be assured of this.

▶ Confidence Intervals Comparing Two Means

Two-Group Design

As before, let $\overline{X}_1$ and $\overline{X}_2$ be the sample means based on units receiving treatments one and two, respectively. Then $\overline{X}_1 - \overline{X}_2$ is a point estimate of the difference in the two treatment means, $\mu_1 - \mu_2$. The standard error of $\overline{X}_1 - \overline{X}_2$ is $\sqrt{\frac{s_p^2}{n_1} + \frac{s_p^2}{n_2}} = s_p\sqrt{\frac{1}{n_1} + \frac{1}{n_2}}$ if $\sigma_1^2 = \sigma_2^2$

and $\sqrt{\frac{s_1^2}{n_1} + \frac{s_2^2}{n_2}}$ if $\sigma_1^2 \neq \sigma_2^2$.

To set a confidence interval on the difference in two treatment means, $\mu_1 - \mu_2$, using the methods outlined here, two conditions must be satisfied. First, the treatments must be *independent* random samples from the population of units receiving treatment 1 and treatment 2. Suppose the units are randomly selected from the population and then randomly assigned to either treatment 1 or treatment 2. The random selection of the units gives us the random samples, and the random assignment of treatment ensures independence. If the units are not randomly selected, then we must rely solely on the random assignment of the treatments to give us a population of all possible samples for the two treatments from these units. Either way, the random assignment of treatments is critical for inference. Second, the responses must either be normally distributed, or the sample size *for each treatment* must be large enough ($n \geq 30$) so that, by the Central Limit Theorem, each estimated treatment mean is approximately normal. Once these conditions are met, the approach we use will depend on whether we believe $\sigma_1^2 = \sigma_2^2$ or $\sigma_1^2 \neq \sigma_2^2$. We will consider these two cases in turn.

First, assume $\sigma_1^2 = \sigma_2^2$. To standardize $\overline{X}_1 - \overline{X}_2$, we take

$$t = \frac{(\overline{X}_1 - \overline{X}_2) - (\mu_1 - \mu_2)}{s_p\sqrt{\frac{1}{n_1} + \frac{1}{n_2}}},$$

which has a t-distribution with $(n_1 + n_2 - 2)$ degrees of freedom. Thus, a $100(1 - \alpha)\%$ confidence interval on $\mu_1 - \mu_2$ is $(\overline{X}_1 - \overline{X}_2) \pm t^* s_p \sqrt{\frac{1}{n_1} + \frac{1}{n_2}}$ where t^* with $(n_1 + n_2 - 2)$ is the tabulated value such that $P(t > t^*) = \frac{\alpha}{2}$.

Next, suppose that $\sigma_1^2 \neq \sigma_2^2$. Standardizing $\overline{X}_1 - \overline{X}_2$, we have $t = \dfrac{(\overline{X}_1 - \overline{X}_2) - (\mu_1 - \mu_2)}{\sqrt{\frac{s_1^2}{n_1} + \frac{s_2^2}{n_2}}}.$

This standardized variable is only approximately distributed as a t-distribution, and the approximation involves a complicated formula for the degrees of freedom. That is,

$$df = \frac{\left(\frac{s_1^2}{n_1} + \frac{s_2^2}{n_2}\right)^2}{\frac{1}{n_1 - 1}\left(\frac{s_1^2}{n_1}\right)^2 + \frac{1}{n_2 - 1}\left(\frac{s_2^2}{n_2}\right)^2}.$$

Typically, computation of these degrees of freedom is built into a calculator or computer software. A $100(1 - \alpha)\%$ confidence interval has the form

$$(\overline{X}_1 - \overline{X}_2) \pm t^* \sqrt{\frac{s_1^2}{n_1} + \frac{s_2^2}{n_2}},$$

where t^* has the degrees of freedom given previously.

Example

Consider the telemarketing example in the previous lesson. Let the subscript H represent the hard-sell approach and the subscript S represent the soft-sell approach. Set an 80% confidence interval on the difference in the treatment means, $\mu_H - \mu_S$.

Solution

Two conditions must be satisfied. Randomly selected members of the sales force were assigned at random to the two treatments, so the first condition is satisfied. We noted earlier that the data consisted of discrete counts so they could not be normally distributed. However, the sample size is 40 for each treatment, allowing us to invoke the Central Limit Theorem.

In the previous lesson, we found $\overline{X}_H - \overline{X}_S = -0.93$ sales. The estimated variance for the hard-sell approach is 1.59 sales^2, and that for the soft-sell approach is 1.93 sales^2. Because these two estimates are close to each other, we assume that $\sigma_1^2 = \sigma_2^2$. Therefore, as we saw earlier, the standard error of $\overline{X}_H - \overline{X}_S$ is $\sqrt{s_P^2\left(\dfrac{1}{n_N} + \dfrac{1}{n_S}\right)} = 0.30$. For an 80% confidence interval, $\alpha = 0.20$. We must find t^*, such that $P(t > t^*) = \dfrac{\alpha}{2} = 0.10$ where we have $(n_1 + n_2 - 2) = 40 + 40 - 2 = 78\ df$. From the t-table in Lesson 12, we look in the row for 78 df and the column for $\alpha = 0.10$ to find $t^* = 1.292$. Therefore, an 80% confidence interval on $\mu_H - \mu_S$ is

$$-0.93 \pm 1.292(0.30)\sqrt{\frac{1}{40} + \frac{1}{40}} \text{ or } -0.93 \pm 0.087.$$

Therefore, we are 80% confident that, on average, the number of sales using the hard-sell approach is between 0.84 and 1.02 less each hour than using the soft-sell approach. Notice that the negative number meant that fewer sales were made using the hard-sell approach because we were estimating $\mu_H - \mu_S$. A positive number would have indicated that the estimated mean for the hard-sell approach was larger than that of the soft-sell approach.

▶ Differences in Two Population Means

For a confidence interval on the difference in two population means to be valid, two conditions must be met. First, samples must be selected randomly and independently from the two populations. Second, for each population, the responses must be normally distributed or the sample size must be sufficiently large to invoke the Central Limit Theorem. If these two conditions are satisfied, the process of establishing the con-

fidence intervals is the same as that used for a two-group design.

Example

Set a 95% confidence interval on the difference in the mean difference in heights for fraternal and identical twins using the data in the previous lesson.

Solution

First, consider the conditions. The twins were recruited and not randomly selected from all fraternal and identical twins. We must assume that these samples are representative of the populations if we are to proceed. We will make this assumption, knowing that it is a potential weakness in our study. Second, only nonnegative values can be observed, and the sample distributions appear skewed. Therefore, the distribution of differences in twin heights is not normal for either fraternal or identical twins. However, the sample size is 30, so we will assume that the Central Limit Theorem can be applied.

The estimated mean difference in the heights for fraternal and identical twins is $\overline{X}_F - \overline{X}_I = 2.4$ cm, and the standard error of this estimate is $\sqrt{\dfrac{s_I^2}{n_I} + \dfrac{s_F^2}{n_F}} = 0.75$ cm, where the subscript F indicates fraternal twins and the subscript I represents identical twins. The degrees of freedom are

$$df = \frac{\left(\dfrac{s_I^2}{n_I} + \dfrac{s_F^2}{n_F}\right)^2}{\dfrac{1}{n_I - 1}\left(\dfrac{s_I^2}{n_I}\right)^2 + \dfrac{1}{n_F - 1}\left(\dfrac{s_F^2}{n_F}\right)^2}$$

$$= \frac{\left(\dfrac{2.10}{30} + \dfrac{14.76}{30}\right)^2}{\dfrac{1}{29}\left(\dfrac{2.10}{30}\right)^2 + \dfrac{1}{29}\left(\dfrac{14.76}{30}\right)^2} = 37.22.$$

To look up the tabulated value, we will round to the nearest integer, 37 in this case. In the row corresponding to 37 df and the column under 0.025 in the t-table, we have 2.024. Based on these values, the 95% confidence interval for $\mu_F - \mu_I$ is

$2.4 \pm 2.024\sqrt{\dfrac{2.10}{30} + \dfrac{14.76}{30}}$ or 2.4 ± 1.518. We estimate that the mean difference in fraternal twins' heights is 2.4 cm greater than the mean difference in identical twins' heights, and we are 95% confident this estimate is within 1.5 cm of the difference in these two population means.

▶ Differences in Two Population Proportions

Sometimes, we want to estimate the difference in two population proportions. For example, we want to estimate "the gender gap," or the difference in the proportions of men and women favoring a particular candidate. Two conditions must be satisfied to use the methods discussed here. First, independent samples are randomly selected from each of the populations. Suppose n_1 and n_2 are the number of observations from populations 1 and 2, respectively. Further, the sample proportions, $\widehat{p}_1$ and $\widehat{p}_2$, are the estimates of population 1 and 2 proportions, $p_1 - p_2$, respectively. The second condition is that $n_1\widehat{p}_1$, $n_1(1 - \widehat{p}_1)$, $n_2\widehat{p}_2$, and $n_2(1 - \widehat{p}_2)$ are all at least 5, and preferably at least 10.

The estimate of the difference in population proportions, $p_1 - p_2$, is $\widehat{p}_1 - \widehat{p}_2$. The standard error of this estimate is $\sqrt{\dfrac{\widehat{p}_1(1 - \widehat{p}_1)}{n_1} + \dfrac{\widehat{p}_2(1 - \widehat{p}_2)}{n_2}}$. Standardizing $\widehat{p}_1 - \widehat{p}_2$, we have

$$z = \frac{(\widehat{p}_1 - \widehat{p}_2) - (p_1 - p_2)}{\sqrt{\dfrac{\widehat{p}_1(1 - \widehat{p}_1)}{n_1} + \dfrac{\widehat{p}_2(1 - \widehat{p}_2)}{n_2}}}.$$

Therefore, the $100(1 - \alpha)$% confidence interval is $(\widehat{p}_1 - \widehat{p}_2) \pm z^{\star}\sqrt{\dfrac{\widehat{p}_1(1 - \widehat{p}_1)}{n_1} + \dfrac{\widehat{p}_2(1 - \widehat{p}_2)}{n_2}}$ where $z^{\star}$ is the tabulated value of z such that $P(z > z^{\star}) = \dfrac{\alpha}{2}$.

▶ Practice

7. Consider the caffeine study in the previous set of practice problems. Set a 95% confidence interval on the mean difference in memory test scores after ingesting caffeine compared to not ingesting caffeine. Is there support for the student's belief that ingesting caffeine while studying improves memory?

▶ In Short

In a two-group design, treatments are randomly assigned to the experimental units. For a two-group design, the methods for setting confidence intervals on the difference in two treatment means were discussed. An important step in this process is determining whether or not the variances of the units under each treatment are equal. The methods are the same when comparing the means of two populations or two population proportions. Although not covered here, the procedures for hypothesis testing have the same extensions as those for confidence intervals.

20 ▶ Analyzing Categorical Data

LESSON SUMMARY

In recent lessons, the responses have been numerical. When working with two treatments or populations, the treatments or populations may be thought of as categorical explanatory variables. In this lesson, we will discuss what to do if the responses, and not just the explanatory variables, are categorical.

▶ Univariate Categorical Data

Univariate categorical data may arise in a variety of settings. In a study of sea turtles, the number of males and females hatched during a season may be recorded. Here two categories exist, male and female, and the data consist of the number of observations falling within each category. A car arriving at an intersection, may continue forward, turn left, turn right, or reverse directions (do a U-turn), resulting in four categories of a univariate response.

The proportion p_i of the population within category $i, i = 1, 2, \ldots, r$, tends to be of primary interest in such studies. As an example, we might hypothesize that the proportion of male and female sea turtles hatched each season is the same (0.50). For the car's movement at a particular intersection, one hypothesis would be that 50% continue forward, 30% turn right, 15% turn left, and 5% do a U-turn. Hypothesis tests are conducted to determine whether or not the observed data are compatible with these hypotheses.

Example

A large grocery store wants to decrease the time needed to check out a customer. One aspect of this is the method by which the customer pays. Payment by cash, check, debit card, or credit card is accepted. The store manager believes that 35% of the transactions are by cash, 5% are by check, 50% are by debit card, and 10% are by credit card. Here the percentages are referring to the number of transactions and not the amount of a transaction. She plans a study to determine whether or not these percentages are correct.

1. Give the response variable and identify the parameters associated with its categories.
2. Set the hypotheses to be tested.

Solution

1. The response variable of interest is a customer's method of payment.
2. We may define the following:

   p_1 = proportion of cash transactions
   p_2 = proportion of check transactions
   p_3 = proportion of debit card transactions
   p_4 = proportion of credit card transactions

It is not important whether we use p_1 to denote the proportion of cash transactions or the proportion of some other type of transaction. However, clearly stating what p_1 represents is important. The set of hypotheses of interest is

$H_0: p_1 = 0.35, p_2 = 0.05, p_3 = 0.50,$
$\quad\quad$ and $p_4 = 0.10$

$H_a:$ not H_0

▶ Practice

A small sandwich shop has six different sandwiches on the menu: turkey, beef, veggie, salami, ham, and pastrami. Unfortunately, the shop has been losing business and, to save money, has to discontinue one of its sandwiches. The owner of the shop believes that 28%, 12%, 20%, 14%, 22%, and 4% of his customers order turkey, beef, veggie, salami, ham, and pastrami, respectively. He plans a study to determine whether or not these percentages are correct.

1. Give the response variable and identify the parameters associated with its categories.
2. Identify the hypotheses to be tested.

▶ Goodness-of-Fit Tests and the χ^2-Distribution

To test null hypotheses of the type just described, we select a random sample from the population of interest and classify each observation into one of the r categories. Let n_i be the number of observations in category $i, i = 1, 2, \ldots, r$. The observed proportion in each category is $\widehat{p}_i = \dfrac{n_i}{n}$; notice that $n_i = n\widehat{p}_i$. Under the null hypothesis, we expect $e_i = n\widehat{p}_i$, observations to be in category i. If the observed n_i and expected e_i counts are equal, the data are fully compatible with the null hypothesis. However, because we expect variability in the sample, we also expect the observed and expected counts to differ to some extent even if the null hypothesis is true. How far apart can they be before we doubt the null hypothesis? The statistical methods that help us make that decision are called goodness-of-fit tests. A *goodness-of-fit* test assesses whether the observed counts in each category are compatible with the hypothesized proportions.

The test statistic for a goodness-of-fit test is $X^2 = \sum_{i=1}^{r} \frac{(n_i - e_i)^2}{e_i}$. Again, notice that if n_i and e_i are equal for $i = 1, 2, \ldots, r$, $X^2 = 0$. As the observed and expected counts become farther apart, X^2 gets larger. We want to reject the null hypothesis when X^2 gets too large. If the null hypothesis is true and the expected counts in each category are not too small, the distribution of X^2 is approximately a χ^2- distribution (pronounced *chi-squared distribution*). Two conditions must be satisfied for this test to be appropriate. First, the sample must be randomly selected from the population of interest. Second, the expected counts in each cell must not be too small. In general, the expected counts should be at least five.

The χ^2- distribution has an area associated with only nonnegative values. The distribution is skewed to the right. The amount of skew, depends on its parameter and the degrees of freedom. Figure 20.1 illustrates how the shape of the distribution changes with degrees of freedom. It illustrates the shape of the distribution for 4, 6, and 10 degrees of freedom. Because we want to reject the null hypothesis if the test statistic X^2 gets too large, the p-value is the area under the curve to the right of the test statistic. Although calculators and computers can be used to find these areas precisely, we will use Table 20.1 to help us approximate it.

The degrees of freedom associated with a goodness-of-fit test are the number of categories minus 1 minus the number of parameters estimated. When no parameters are estimated, the degrees of freedom are $r - 1$, or the number of categories minus one. As in earlier tests of hypotheses, we reject the null hypothesis if the p-value gets too small. Typically, a value less than 0.05 is significant, and a value less than 0.01 is highly significant.

This type of hypothesis test differs from the others we have considered. We are unable to have the research hypothesis as the alternative hypothesis. Here, we want to show that the null hypothesis is true. However, at best, our conclusion will be that the data are consistent or compatible with the null hypothesis. If we reject the null hypothesis, we have strong evidence that it is not true.

We should make one final note. If only two categories exist, the χ^2-test is an alternative to a two-sided test for a population proportion. The two often provide similar but not exactly the same results. If a one-sided test of a proportion is needed, the χ^2-test as presented here cannot be used.

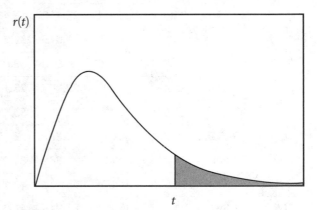

Figure 20.2

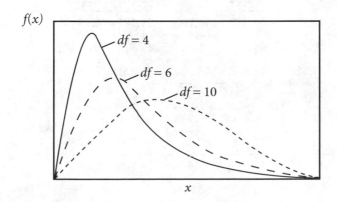

Figure 20.1

Table 20.1 Selected right tail areas

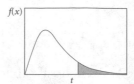

df	0.3000	0.2000	0.1500	0.1000	0.0500	0.0250	0.0100	0.0050	0.0010	0.000
1	1.074	1.642	2.072	2.706	3.841	5.024	6.635	7.879	10.828	12.116
2	2.408	3.219	3.794	4.605	5.991	7.378	9.210	10.597	13.816	15.202
3	3.665	4.642	5.317	6.251	7.815	9.348	11.345	12.838	16.266	17.730
4	4.878	5.989	6.745	7.779	9.488	11.143	13.277	14.860	18.467	19.997
5	6.064	7.289	8.115	9.236	11.070	12.833	15.086	16.750	20.515	22.105
6	7.231	8.558	9.446	10.645	12.592	14.449	16.812	18.548	22.458	24.103
7	8.383	9.803	10.748	12.017	14.067	16.013	18.475	20.278	24.322	26.018
8	9.524	11.030	12.027	13.362	15.507	17.535	20.090	21.955	26.124	27.868
9	10.656	12.242	13.288	14.684	16.919	19.023	21.666	23.589	27.877	29.666
10	11.781	13.442	14.534	15.987	18.307	20.483	23.209	25.188	29.588	31.420
11	12.899	14.631	15.767	17.275	19.675	21.920	24.725	26.757	31.264	33.137
12	14.011	15.812	16.989	18.549	21.026	23.337	26.217	28.300	32.909	34.821
13	15.119	16.985	18.202	19.812	22.362	24.736	27.688	29.819	34.528	36.478
14	16.222	18.151	19.406	21.064	23.685	26.119	29.141	31.319	36.123	38.109
15	17.322	19.311	20.603	22.307	24.996	27.488	30.578	32.801	37.697	39.719
16	18.418	20.465	21.793	23.542	26.296	28.845	32.000	34.267	39.252	41.308
17	19.511	21.615	22.977	24.769	27.587	30.191	33.409	35.718	40.790	42.879
18	20.601	22.760	24.155	25.989	28.869	31.526	34.805	37.156	42.312	44.434
19	21.689	23.900	25.329	27.204	30.144	32.852	36.191	38.582	43.820	45.973
20	22.775	25.038	26.498	28.412	31.410	34.170	37.566	39.997	45.315	47.498
21	23.858	26.171	27.662	29.615	32.671	35.479	38.932	41.401	46.797	49.011
22	24.939	27.301	28.822	30.813	33.924	36.781	40.289	42.796	48.268	50.511
23	26.018	28.429	29.979	32.007	35.172	38.076	41.638	44.181	49.728	52.000
24	27.096	29.553	31.132	33.196	36.415	39.364	42.980	45.559	51.179	53.479
25	28.172	30.675	32.282	34.382	37.652	40.646	44.314	46.928	52.620	54.947
26	29.246	31.795	33.429	35.563	38.885	41.923	45.642	48.290	54.052	56.407
27	30.319	32.912	34.574	36.741	40.113	43.195	46.963	49.645	55.476	57.858
28	31.391	34.027	35.715	37.916	41.337	44.461	48.278	50.993	56.892	59.300
29	32.461	35.139	36.854	39.087	42.557	45.722	49.588	52.336	58.301	60.735
30	33.530	36.250	37.990	40.256	43.773	46.979	50.892	53.672	59.703	62.162
40	44.165	47.269	49.244	51.805	55.758	59.342	63.691	66.766	73.402	76.095
50	54.723	58.164	60.346	63.167	67.505	71.420	76.154	79.490	86.661	89.561

Example

Find the p-values associated with the following values of the test statistic, X^2:

1. $X^2 = 3.841$, $df = 1$
2. $X^2 = 3.841$, $df = 2$
3. $X^2 = 3.841$, $df = 5$
4. $X^2 = 50$, $df = 20$

Solution

1. We look across the row with 1 degree of freedom for the values closest to 3.841. The value 3.841 is in the 0.05 column. Thus, the p-value is 0.05.

2. This time, we look across the row for 2 degrees of freedom. The values 3.794 and 4.605 are in the 0.15 and 0.10 columns, respectively. This means that the p-value is between 0.10 and 0.15.

3. Looking across the row for 5 degrees of freedom, the smallest value is 6.064 in the 0.30 column. Therefore, the p-value is greater than 0.30.

4. In the row for 20 degrees of freedom, the largest value is 47.498 in the 0.0005 column. Thus, the p-value is less than 0.0005.

▶ Practice

Find the p-values associated with the following values of the test statistic, X^2:

1. $X^2 = 9.348$, $df = 3$
2. $X^2 = 9.348$, $df = 5$
3. $X^2 = 35$, $df = 11$
4. $X^2 = 9.348$, $df = 10$

Example

The store manager randomly selected 200 transactions from those made during the past week and recorded the method by which the customer is paid. Of the 200 purchases, 62 were paid by cash, 8 were paid by check, 112 were paid by debit card, and 18 were paid by credit card.

1. Find the expected counts within each payment category.

2. Determine the value of the test statistic.

3. Find the p-value associated with the test statistic.

4. Determine whether or not to reject the null hypothesis and state the conclusion in the context of the problem.

Solution

1. The observed and expected counts are shown in Table 20.2.

Table 20.2 Observed and expected counts

PAYMENT METHOD	NUMBER OBSERVED	NUMBER EXPECTED	$\frac{(n_i - e_i)^2}{e_i}$
Cash	62	$200 \times 0.35 = 70$	$\frac{(62 - 70)^2}{70} = 0.1943$
Check	8	$200 \times 0.05 = 10$	0.4000
Debit Card	112	$200 \times 0.50 = 100$	1.4400
Credit Card	18	$200 \times 0.10 = 20$	0.2000
Totals	200	200	2.9543

2. Organizing the information as in the table in problem 1, generally simplifies computations. Recall the test statistic is $X^2 = \sum_{i=1}^{r} \frac{(n_i - e_i)^2}{e_i}$. Each term in this summation is found in the $\frac{(n_i - e_i)^2}{e_i}$ column. The column total is $X^2 = 2.9543$.

3. The degrees of freedom associated with the test are $4 - 1 = 3$ because there are four categories and no parameters were estimated. The smallest value in the row associated with 3 degrees of freedom is 3.665, which is in the column labeled 0.30. Thus, the p-value is greater than 0.30.

4. The p-value is greater than 0.30 and would also be larger than any commonly used significance level. Therefore, we would not reject the null hypothesis that the percentages of cash, check, debit card, and credit card transactions are as the store manager has hypothesized.

▶ Practice

The sandwich shop owner randomly selected 150 of his customers that purchased a sandwich within the last month and recorded the type of sandwich the customer ordered. Of the 150 orders, 45 ordered turkey, 14 ordered beef, 33 ordered veggie, 16 ordered salami, 37 ordered ham, and 5 ordered pastrami.

5. Find the expected counts within each sandwich category.

6. Determine the value of the test statistic.

7. Find the p-value associated with the test statistic.

8. Determine whether or not to reject the null hypothesis and state the conclusion in the context of the problem.

▶ Tests of Homogeneity

Suppose we want to know whether the proportions within each of the r categories of a response variable are the same for each of c populations. To investigate this question, independent samples are taken from each of the r populations. The data can be arranged in an $r \times c$ table, called a *contingency table*. The general form of a contingency table is shown here in Table 20.3.

Table 20.3 Contingency table

CATEGORY	POPULATION 1	2	.	.	c	TOTALS
1	n_{11}	n_{12}	.	.	n_{1c}	$n_{1.}$
2	n_{21}	n_{22}	.	.	n_{2c}	$n_{2.}$
.						
.						
r	n_{r1}	n_{r2}	.	.	n_{rc}	$n_{r.}$
Totals	$n_{.1}$	$n_{.2}$	.	.	$n_{.c}$	$n_{..}$

The first subscript on a count in Table 20.3 corresponds to the row and the second to the column. Row, column, and overall totals are also in the table. Periods are used to show which variable summation was over. For example, $n_{1.}$ is the total of the first row. The total of the second column is $n_{.2}$, and $n_{..}$ is the overall total.

The null hypothesis is that the proportion in each category is the same for all populations, i.e., $H_0: p_{1j} = p_1, p_{2j} = p_2, \ldots, p_{rj} = p_r, j = 1, 2,, \ldots, c$. The alternative hypothesis is H_a: not H_0. To estimate the proportion in the ith category common to all populations, we use the sample proportion in the ith category across all populations, $\widehat{p}_i = \dfrac{n_{i.}}{n_{..}}$. The expected count in the ith category from the jth population is the size of the sample from that population times the sample proportion in the ith category; that is, $e_{ij} = n_j \widehat{p}_i = \dfrac{n_{i.} n_{.j}}{n_{..}}$ or the row total times the column total divided by the overall total. The test statistic is

$$X^2 = \sum_{i=1}^{r} \sum_{j=1}^{c} \frac{(n_{ij} - e_{ij})^2}{e_{ij}}.$$ If each population was sampled randomly and all expected counts are large enough (≥ 5), the test statistic has an approximate χ^2- distribution with $(r-1)(c-1)$ degrees of freedom *provided* the null hypothesis is true. The p-value is the probability that a randomly selected value from the χ^2- distribution exceeds the test statistic. As with other hypothesis tests, we reject the null hypothesis if the p-value is too small.

Example

A researcher wanted to determine whether the age distribution is the same for renters and for home owners in a large city. He selected a random sample of size 100 from all renters and recorded their ages in categories. Similarly, he selected a random sample of size 100 from all home owners and recorded their ages in categories. The data are presented in Table 20.4.

1. State the null and alternative hypotheses of interest to this researcher.
2. Find the expected counts.
3. Verify the conditions for the test are satisfied and, if so, find the value of the test statistic
4. Find the p-value.
5. Decide whether or not to reject the null hypothesis.
6. State the conclusions in the context of the problem.

Table 20.4 Age of renters and home owners

AGE CATEGORY	RENTERS	HOME OWNERS	TOTALS
18–25	22	3	25
26–40	36	25	61
41–55	25	36	61
56–65	7	17	24
Over 65	10	19	29
Totals	100	100	200

Solution

1. $H_0: p_{1j} = p_1, p_{2j} = p_2, \ldots, p_{rj} = p_r, j = R, H$, where R stands for renters and H represents home owners. The alternative hypothesis is H_a: not H_0.

2. The expected counts are in Table 20.5.

Table 20.5 Expected counts

AGE CATEGORY	RENTERS	HOME OWNERS	TOTALS
18–25	$\frac{25 \times 100}{200} = 12.5$	12.5	25
26–40	30.5	30.5	61
41–55	30.5	30.5	61
56–65	12	12	24
Over 65	14.5	14.5	29
Totals	100	100	200

Notice that the expected counts do not have to be, and are often not, whole numbers. They should *not* be rounded to whole numbers.

3. A random sample was selected from each population. The smallest expected count was 12, which is greater than the minimum of 5 needed for the test statistic to have an approximate χ^2- distribution. Thus, the conditions for the test are satisfied. The value of the test statistic is

$$X^2 = \sum_{i=1}^{r} \sum_{j=1}^{c} \frac{(n_{ij} - e_{ij})^2}{e_{ij}} = 25.3670.$$

4. The degrees of freedom associated with the test statistic are $(r - 1)(c - 1) = 4 \times 1 = 4$. The largest number on the row with 4 degrees of freedom is 19.997, which is in the 0.0005 column. Therefore, the p-value is less than 0.0005.

5. Because the p-value is so very small, the evidence against the null hypothesis is very strong. Thus, we reject it in favor of the alternative hypothesis.

6. Evidence exists that the proportions in each age category is not the same for home owners and renters.

▶ Practice

A researcher wanted to know whether men and women drive the same types of vehicles in the researcher's home state. She randomly selected 125 males and 125 females from all vehicle owners in her state and recorded what type of vehicle each drives. The data are presented in Table 20.6.

Table 20.6 Vehicle of choice for men and women

TYPE OF VEHICLE	MALE	FEMALE	TOTALS
Trucks	44	16	60
Vans	8	27	35
Sport Utility Vehicles (SUVs)	23	32	55
Convertibles	9	8	17
Sedans	30	33	63
Other	11	9	20
Totals	125	125	250

9. State the null and alternative hypotheses of interest to this researcher.
10. Find the expected counts.
11. Verify whether or not the conditions for the test are satisfied and, if so, find the value of the test statistic.
12. Find the p-value.
13. Decide whether or not to reject the null hypothesis.
14. State the conclusions in the context of the problem.

▶ Tests of Independence

Sometimes, data on two categorical variables can be collected in one sample. For example, instead of sampling renters and home owners separately in the previous example, we could have taken one sample and asked each study participant whether he or she is a renter or a home owner and which age category he or she is in. The data could have been presented in a table as in the example. The only difference is the manner in which the data were collected. The null hypothesis is that the two variables are independent of one another, and the alternative is that they are not. The expected counts are then computed as in the tests for homogeneity. The conditions that must be satisfied are that the sample was randomly selected and that the expected counts are large enough, at least five in each cell. If these are satisfied, the test statistic,

$$X^2 = \sum_{i=1}^{r} \sum_{j=1}^{c} \frac{(n_{ij} - e_{ij})^2}{e_{ij}},$$

has an approximate χ^2-distribution with $(r-1)(c-1)$ degrees of freedom if the null hypothesis is true. The p-value is the probability that a randomly selected observation from the χ^2-distribution is greater than the test statistic. If this value is less than the specified significance level, the null hypothesis is rejected; otherwise, the null is not rejected.

Example

A company wanted to assess the success of its television advertising campaign for a new product. They hired a pollster to find out whether those who saw the ad were more likely to have purchased the new product than those who had not. The pollster took a sample of 250 adults in the viewing area where the ad aired. Each study participant was asked whether he or she had seen the ad and whether he or she had purchased the new product. The results are presented in Table 20.7.

Table 20.7 The effect of the ad on sales

PURCHASED PRODUCT	SAW THE TELEVISION AD		
	YES	NO	TOTALS
Yes	20	45	65
No	50	135	185
Totals	70	180	250

1. State the null and alternative hypotheses of interest to the company.
2. Find the expected counts.
3. Verify the conditions for the test and, if satisfied, find the test statistic.
4. Find the p-value.
5. Decide whether or not to reject the null hypothesis.
6. State your conclusion. Be sure it is in the context of the problem.

Solution

1. H_0: Having viewed the ad is independent of whether or not a person purchased the product. H_a: Having viewed the ad is not independent of whether or not a person purchased the product
2. The expected counts are in Table 20.8.

Table 20.8

PURCHASED PRODUCT	SAW THE TELEVISION AD		
	YES	NO	TOTALS
Yes	$\frac{65 \times 70}{250} = 18.2$	46.8	65
No	51.8	133.2	185
Totals	70	180	250

3. The sample was randomly selected from the population that had the opportunity to see the ad, and all expected counts exceed five. Thus the conditions for the test are satisfied. The test statistic is then

$$X^2 = \sum_{i=1}^{r} \sum_{j=1}^{c} \frac{(n_{ij} - e_{ij})^2}{e_{ij}} = 0.3341.$$

4. If the null hypothesis is true, the test statistic has an approximate χ^2-distribution with $(r-1)(c-1) = 1 \times 1 = 1$ degree of freedom. The smallest value in the row of the χ^2-table corresponding to one degree of freedom is 1.074 in the 0.3 column. Thus, the p-value is greater than 0.30.

5. The p-value is large, indicating that data such as what was observed are not at all unusual if the null hypothesis is true. Therefore, we would not reject the null hypothesis.

6. There is not sufficient evidence to reject the hypothesis that seeing the ad is independent of whether or not the new product was purchased. This would be frustrating information for the company's management. The lack of a significant relationship indicates that no sufficient evidence indicates that people were more likely to purchase the product after seeing the television ad. The company may be looking for a new advertisement firm!

▶ **Practice**

15. The student government wanted to know whether or not the freshmen who live on campus develop more school spirit by the end of the academic year than those who do not live on campus. They commissioned a poll to find out. The pollster randomly selected freshmen to participate in the study. Each selected freshman was asked whether or not he or she lives on campus and asked whether he or she considers him- or herself to have no school spirit, some school spirit, or a lot of school spirit. The results are shown in Table 20.9.

Table 20.9 Freshman with school spirit

SCHOOL SPIRIT	LIVE ON CAMPUS?	
	Yes	No
A lot	67	12
Some	42	32
None	6	21

Is there statistical evidence to support an association between whether a freshman lives on campus and the amount of his or her school spirit?

▶ In Short

Categorical data lead to counts within each category. χ^2-tests are suitable for testing hypotheses about these data. When working with univariate categorical data, one can test whether the population proportions in each category are some set of specified values. If the same univariate categorical variable is observed in independent samples from two or more populations, one can test whether the proportions in each category are the same for all populations. If two different categorical variables are observed in one sample, the test concerns whether or not the two variables are independent.

Posttest

If you have completed all 20 lessons in this book, then you are ready to take the posttest to measure your progress. The posttest has 50 multiple-choice questions covering the topics you studied in this book. Although the format of the posttest is similar to that of the pretest, the questions are different.

Take as much time as you need to complete the posttest. When you are finished, check your answers with the answer key that follows. Once you know your score on the posttest, compare the results with the pretest. If you score better on the posttest than you did on the pretest, congratulations! You have profited from your hard work. At this point, you should look at the questions you missed, if any. Do you know why you missed the question, or do you need to go back to the lesson and review the concept?

If your score on the posttest doesn't show much improvement, take a second look at the questions you missed. Did you miss a question because of an error you made? If you can figure out why you missed the problem, then you understand the concept and simply need to concentrate more on accuracy when taking a test. If you missed a question because you did not know how to work the problem, go back to the lesson and spend more time working that type of problem. Take the time to understand basic statistics thoroughly. You need a solid foundation in statistics if you plan to use this information or progress to a higher level of statistics. Whatever your score on this posttest, keep this book for review and future reference.

ANSWER SHEET

1. ⓐ ⓑ ⓒ ⓓ
2. ⓐ ⓑ ⓒ ⓓ
3. ⓐ ⓑ ⓒ ⓓ
4. ⓐ ⓑ ⓒ ⓓ
5. ⓐ ⓑ ⓒ ⓓ
6. ⓐ ⓑ ⓒ ⓓ
7. ⓐ ⓑ ⓒ ⓓ
8. ⓐ ⓑ ⓒ ⓓ
9. ⓐ ⓑ ⓒ ⓓ
10. ⓐ ⓑ ⓒ ⓓ
11. ⓐ ⓑ ⓒ ⓓ
12. ⓐ ⓑ ⓒ ⓓ
13. ⓐ ⓑ ⓒ ⓓ
14. ⓐ ⓑ ⓒ ⓓ
15. ⓐ ⓑ ⓒ ⓓ
16. ⓐ ⓑ ⓒ ⓓ
17. ⓐ ⓑ ⓒ ⓓ

18. ⓐ ⓑ ⓒ ⓓ
19. ⓐ ⓑ ⓒ ⓓ
20. ⓐ ⓑ ⓒ ⓓ
21. ⓐ ⓑ ⓒ ⓓ
22. ⓐ ⓑ ⓒ ⓓ
23. ⓐ ⓑ ⓒ ⓓ
24. ⓐ ⓑ ⓒ ⓓ
25. ⓐ ⓑ ⓒ ⓓ
26. ⓐ ⓑ ⓒ ⓓ
27. ⓐ ⓑ ⓒ ⓓ
28. ⓐ ⓑ ⓒ ⓓ
29. ⓐ ⓑ ⓒ ⓓ
30. ⓐ ⓑ ⓒ ⓓ
31. ⓐ ⓑ ⓒ ⓓ
32. ⓐ ⓑ ⓒ ⓓ
33. ⓐ ⓑ ⓒ ⓓ
34. ⓐ ⓑ ⓒ ⓓ

35. ⓐ ⓑ ⓒ ⓓ
36. ⓐ ⓑ ⓒ ⓓ
37. ⓐ ⓑ ⓒ ⓓ
38. ⓐ ⓑ ⓒ ⓓ
39. ⓐ ⓑ ⓒ ⓓ
40. ⓐ ⓑ ⓒ ⓓ
41. ⓐ ⓑ ⓒ ⓓ
42. ⓐ ⓑ ⓒ ⓓ
43. ⓐ ⓑ ⓒ ⓓ
44. ⓐ ⓑ ⓒ ⓓ
45. ⓐ ⓑ ⓒ ⓓ
46. ⓐ ⓑ ⓒ ⓓ
47 ⓐ ⓑ ⓒ ⓓ
48. ⓐ ⓑ ⓒ ⓓ
49. ⓐ ⓑ ⓒ ⓓ
50. ⓐ ⓑ ⓒ ⓓ

▶ Posttest

1. The number of students eating a meal in the school cafeteria is the variable of interest. What type of variable is being observed?

a. a categorical variable

b. a continuous variable

c. a discrete variable

d. a explanatory variable

2. A study was conducted to compare the strength of four paper towel brands. A roll of each of the paper towel brands was purchased in a local store. Ten sheets were taken from each roll and tested for strength. What type of study is this?

a. an experiment with a broad scope of inference

b. an experiment with a narrow scope of inference

c. a sample survey

d. an observational study

3. Random-digit dialing was used to select households in a particular state. An adult in each household contacted was asked whether anyone in the household had been imprisoned within the past ten years. A critic of the poll said that the results were biased because people living in households where someone has been imprisoned in the past ten years might be embarrassed and respond no. As a consequence, the estimated percentage of households in which someone had been imprisoned in the past ten years would be biased downward. What type of bias was the critic concerned about?

a. a measurement bias

b. a nonresponse bias

c. a response bias

d. a selection bias

4. A company has a newly developed sugar substitute that is believed to taste better than the sugar substitute it currently makes. To find out if this is correct, 40 students were randomly selected from the students at a nearby large university. Each selected student was asked to taste the two substitutes in random order. A glass of water was drunk between the two tastings to eliminate lingering effects from tasting the first substitute. Each student was asked to identify which sugar substitute tasted better. What is the population of interest and what are the response and explanatory variables?

a. The population is all students at the large university, the response variable is the identification of the better tasting sugar substitute, and the explanatory variable is the type of sugar substitute.

b. The population is all students at the large university, the response variable is the type of sugar substitute, and the explanatory variable is the identification of the better tasting sugar substitute.

c. The population is the two sugar substitutes, the response variable is the identification of the better tasting sugar substitute, and the explanatory variable is the type of sugar substitute.

d. The population is the two sugar substitutes, the response variable is the type of sugar substitute, and the explanatory variable is the identification of the better tasting sugar substitute.

For problems 5 and 6, consider the following twelve data points: 8, 10, 8, 16, 14, 13, 7, 12, 9, 11, 10, and 14.

5. What is the median of these data?
   a. 10
   b. 10.5
   c. 11
   d. The median is not unique.

6. What is the interquartile range of these data?
   a. 4
   b. 5
   c. 6
   d. 9

7. What does the line in the middle of the box in a boxplot represent?
   a. the first quartile
   b. the median
   c. the third quartile
   d. the mean

8. A random variable has a mean of 10 and a standard deviation of 5. After standardizing the random variable, what is its mean and variance?
   a. The mean is 0, and the standard deviation is 1.
   b. The mean is 0, and the standard deviation is 5.
   c. The mean is 10, and the standard deviation is 1.
   d. The mean is 10, and the standard deviation is 5.

Use the following information for problems 9 and 10. In a large city, the proportion of households having a dog is 0.4. The proportion of households having a cat is 0.3. The proportion of households having both a dog and a cat is 0.15.

9. What proportion of households has a dog or a cat?
   a. 0.45
   b. 0.55
   c. 0.70
   d. 0.85

10. What proportion of households has neither a dog nor a cat?
   a. 0.15
   b. 0.30
   c. 0.45
   d. 0.85

Use the following information for problems 11, 12, and 13. The students in a small high school were surveyed. Each student was asked whether he or she consistently exceeded the speed limit. This information and the gender of the student was recorded as follows:

	EXCEED THE SPEED LIMIT?		
GENDER	**YES**	**NO**	**TOTALS**
Female	52	58	110
Male	76	39	115
Totals	128	97	225

11. What is the probability that a randomly selected student is a female who does not consistently exceed the speed limit?

a. $\dfrac{58}{110}$

b. $\dfrac{58}{225}$

c. $\dfrac{97}{225}$

d. $\dfrac{110}{225}$

12. What is the probability that a randomly selected student is a male, given that the person consistently exceeds the speed limit?

a. $\dfrac{76}{128}$

b. $\dfrac{76}{225}$

c. $\dfrac{115}{225}$

d. $\dfrac{128}{225}$

13. Is the use of a safety belt independent of gender?

a. no, because the probability of male does not equal the probability of male, given the student consistently exceeds the speed limit

b. no, because the number of males who consistently exceed the speed limit is not equal to the number of males who consistently exceed the speed limit

c. yes, because the sample was randomly selected

d. yes, because the probability that a randomly selected male consistently exceeds the speed limit is greater than the probability that a randomly selected female consistently speeds

Use the following information for problems 14 and 15: 0.05% of the parts produced in a manufacturing process are defective. Each part is tested for defects. If

the part is defective, the test will indicate it is defective 98% of the time. If the part is not defective, the test will indicate it is defective 1% of the time.

14. What is the probability that a randomly selected part from this population tests defective?

a. 0.0095

b. 0.01485

c. 0.0585

d. 0.98

15. A part is randomly selected from this population and tested. It tests defective. Which of the following best represents the probability that the part is defective?

a. 0.0490

b. 0.1624

c. 0.8376

d. 0.98

Use the following information for problems 16, 17, and 18. A farmer has determined the probability that a hen lays an egg on any given day is 0.9. Whether she lays or not is independent from day to day.

16. The hen laid an egg today. What is the probability that she will NOT lay an egg tomorrow?

a. 0.09

b. 0.1

c. 0.8

d. 0.9

17. Which of the following is closest to the probability that the hen will NOT lay an egg at least one of the next three days?

a. 0.001

b. 0.271

c. 0.729

d. 0.999

18. What is the probability that the hen will lay an egg four days and then fail to lay one on the fifth day?

    a. 0.00009

    b. 0.06561

    c. 0.09

    d. 0.6561

19. Ben is due to work at 8 A.M. He is equally likely to get to work any time between 7:50 and 8:05. Which of the following is closest to the probability Ben will be late on a randomly selected day?

    a. 0.25

    b. 0.33

    c. 0.5

    d. 0.67

20. Let z be a standard normal random variable. Find the probability that a randomly selected value of z is between -1.6 and 1.2.

    a. 0.0548

    b. 0.1151

    c. 0.8301

    d. 0.8849

21. Let z be a standard normal random variable. Find z^* such that the probability that a randomly selected value of z is less than z^* is 0.2.

    a. -0.84

    b. 0.4207

    c. 0.5793

    d. 0.84

22. Let X be a normal random variable with mean 16 and standard deviation 2. What is the probability that a randomly selected value of X is between 12 and 20?

    a. 0.32

    b. 0.68

    c. 0.95

    d. 0.997

23. A random sample of size 16 is selected from a population that is normally distributed with a mean of 20 and a standard deviation of 8. What is the sampling distribution of the sample mean?

    a. normal with a mean of 0 and a standard deviation of 1

    b. normal with a mean of 20 and a standard of 0.5

    c. normal with a mean of 20 and a standard deviation of 2

    d. normal with a mean of 20 and a standard deviation of 8

24. Find t^* such that the probability that a randomly selected observation from a t-distribution with 12 degrees of freedom is greater than t^* is 0.01.

    a. -3.955

    b. -2.681

    c. 2.681

    d. 3.055

25. A researcher decides to study how much a sparrow eats during a day. He believes that if he takes a large enough sample that he will be able to say that the sample mean amount of food consumed daily by the sparrows he observes is close to the population mean. Is he correct?

a. no, one can never be sure that the sample mean is close to the population mean

b. yes, by the Central Limit Theorem, the sample mean will be equal to the population mean if $n > 30$

c. yes, by the Central Limit Theorem, the sample mean will be approximately normally distributed, and the mean of the sampling distribution will be the population mean

d. yes, by the Law of Large Numbers, as the sample size increases, the sample mean will get close to the population mean

26. A poll was conducted to determine what percentage of the registered voters in a swing state favored the incumbent president in a close race for re-election. The results were that 51% of the registered voters polled were in favor of the second term with a margin of error of 0.03. What does this mean?

a. There is only a 3% chance that 51% of the registered voters do not favor the incumbent president for re-election.

b. The estimated percentage of 51% is sure to be within 3% of the true percentage favoring the incumbent president for re-election.

c. With 95% confidence, the estimated percentage of 51% is within 3% of the true percentage favoring the incumbent president for re-election.

d. With 97% confidence, the estimated percentage of 51% is equal to the true percentage favoring the incumbent president for re-election.

27. A blood bank in a large city noticed that the percentages of each blood type from donors seemed to differ from the national averages. Wanting to be sure that they kept adequate supplies for the blood types in their community, they decided to conduct a survey. They randomly selected blocks within the city. A nurse was sent to each selected block and recorded the blood type of each person living in that block. The information from all selected blocks was combined to estimate the percentages of each blood type for the city. What type of sampling plan is this?

a. cluster sampling

b. simple random sampling

c. stratified random sampling

d. systematic random sampling

28. A researcher was interested in estimating the percentage of two-income households in a state. To do this, she took a random sample of households within each county and determined whether or not each selected household had two incomes. She then combined the information from the counties to get an estimate for the state. What type of sampling plan is this?

a. cluster sampling

b. simple random sampling

c. stratified random sampling

d. systematic random sampling

29. A student set a 90% confidence interval on the mean time it took him to get from his home to the band practice field and found it to be 12.4 to 15.6 minutes. Which of the following is an appropriate interpretation of this interval?
a. Ninety percent of the time, the student will take between 12.4 and 15.6 minutes to get to the practice field.
b. There is a 90% probability that the mean time it takes the student to get to the practice field is between 12.4 and 15.6 minutes.
c. We are 90% confident that the sample mean time that the student took to get to the practice field is between 12.4 and 15.6 minutes.
d. We are 90% confident that the mean time that the student takes to get to the practice field is between 12.4 and 15.6 minutes.

30. A large university wanted to know whether the students would favor a $5 increase in student fees to fund popular bands to give free student concerts on campus. A random sample of 250 students was selected. Each selected student was asked whether he or she favored the increase in student fees for this purpose. Of those sampled, 57% favored the increase. Which of the following would be used to set a 90% confidence interval on the proportion of all students at this university who would favor such an increase?

a. $0.57 \pm 1.28 \dfrac{\sqrt{0.57(1 - 0.50)}}{250}$

b. $0.57 \pm 1.28 \sqrt{\dfrac{0.57(1 - 0.57)}{250}}$

c. $0.57 \pm 1.645 \sqrt{0.57(1 - 0.57)}$

d. $0.57 \pm 1.645 \sqrt{\dfrac{0.57(1 - 0.57)}{250}}$

31. A homeowner thinks she may need a new roof. Because of costs, she wants to be sure a new roof is really needed before having the current one replaced. What is the homeowner's null hypothesis and what would be a type I error?
a. H_0: A new roof is not needed. A type I error would occur if she had a new roof put on when it was not needed.
b. H_0: A new roof is not needed. A type I error would occur if she did not have a new roof put on when it was needed.
c. H_0: A new roof is needed. A type I error would occur if she had a new roof put on when it was not needed.
d. H_0: A new roof is needed. A type I error would occur if she did not have a new roof put on when it was needed.

32. The mean value of homes in a large city was reported to be $157,000. The local chamber of commerce thought the mean value of homes was higher than this reported value. Let μ be the mean value of home in the city, and let $\overline{X}$ be the mean value of homes in a randomly selected sample of homes in the city. What is the appropriate set of hypotheses for the chamber of commerce to test?
a. H_0: $\overline{X} = \$157,000$; H_a: $\overline{X} > \$157,000$
b. H_0: $\mu = \$157,000$; H_a: $\mu > \$157,000$
c. H_0: $\mu > \$157,000$; H_a: $\mu = \$157,000$
d. H_0: $\mu \neq \$157,000$; H_a: $\mu = \$157,000$

33. Nationally, the percentage of people with asthma has been reported to be 6%. A researcher believes that a lower percentage than that has asthma in his region. To test this assumption, he selects a random sample of people in the region and determines whether each person has asthma. Of 200 people surveyed, 9 had asthma. What is the appropriate test statistic to test the researcher's hypothesis?

a. $z_T = \dfrac{0.045 - 0.06}{\sqrt{\dfrac{0.045(1 - 0.045)}{200}}}$

b. $z_T = \dfrac{0.045 - 0.06}{\sqrt{\dfrac{0.06(1 - 0.06)}{200}}}$

c. $z_T = \dfrac{0.045 - 0.06}{\dfrac{\sqrt{0.045(1 - 0.045)}}{200}}$

d. $z_T = \dfrac{0.045 - 0.06}{\dfrac{\sqrt{0.06(1 - 0.06)}}{200}}$

34. A statistician was testing the following set of hypotheses: $H_0: p = 0.1$ versus $H_a: p \neq 0.1$. Using a random sample of size 230, she found $z_T = 1.37$. What is the p-value associated with this test?

a. 0.0426
b. 0.0853
c. 0.1706
d. 0.9147

35. A statistician conducted a hypothesis test and found the p-value to be 0.06. Using a 10% level of significance, what conclusion should she make?

a. Accept the null hypothesis.
b. Do not reject the null hypothesis.
c. Reject the alternative hypothesis.
d. Reject the null hypothesis.

Use the following information for problems 36 and 37. An ichthyologist (one who studies fish) wanted to determine the average length of great white sharks in a region that he was studying. He randomly selected 31 of them and measured the length of each. The sample mean was 15.2 feet, and the sample standard deviation was 1.9 feet.

36. What is the appropriate multiplier to use in setting a 90% confidence interval on the mean length of the great white shark in this region?

a. 1.645
b. 1.697
c. 2.040
d. 2.042

37. Given the proper multiplier, which of the following represents a 90% confidence interval on the mean height of trees in this area?

a. $15.2 \pm \textit{multiplier} \times 1.9$

b. $15.2 \pm \textit{multiplier} \times \sqrt{\dfrac{1.9}{31}}$

c. $15.2 \pm \textit{multiplier} \times \dfrac{1.9}{\sqrt{30}}$

d. $15.2 \pm \textit{multiplier} \times \dfrac{1.9}{31}$

38. A car manufacturer wanted to be sure that her cars got at least the average 32 miles/gallon advertised. She selected a random sample of 30 cars and determined the miles/gallon for each. The sample mean was 32.5 miles/gallon, and the sample standard deviation was 2.1 miles/gallon. What is the appropriate test statistic to test the hypothesis that the manufacturer is interested in?

a. $t_T = \dfrac{32 - 32.5}{\dfrac{2.1}{\sqrt{30}}}$

b. $t_T = \dfrac{32 - 32.5}{2.1}$

c. $t_T = \dfrac{32.5 - 32}{\dfrac{2.1}{\sqrt{31}}}$

d. $t_T = \dfrac{32.5 - 32}{2.1}$

39. A statistician conducts a test of the following set of hypotheses: $H_0: \mu = 29$ versus the alternative $\mu < 29$. Based on a random sample of 38, he found the value of the test statistic to be 1.85. What is the p-value associated with the test?

a. $0.0125 < p < 0.025$
b. $0.025 < p < 0.05$
c. $0.05 < p < 0.10$
d. $p = 0.0322$

40. A botanist believes that she has developed a new fertilizer that promotes growth in flowers better than the standard fertilizer. She randomly assigns half of her flower plants to the new fertilizer and half to the standard one. The plants are each properly cared for throughout the growing season, and the growth of each plant is recorded at the end of the season. Statistically, what does the researcher want to do?

a. Compare two treatment means using a matched-pairs design.
b. Compare two treatment means using a two-group design.
c. Compare the means from two populations.
d. Compare the means of two samples from the same population.

41. In a two-group design, 37 observations were taken under the first treatment, and 33 were taken under the second treatment. The sample variance under the first treatment was 9.3 and under the second was 9.8. Believing that both of these are estimates of a common variance, the statistician wants to obtain an estimate of this common variance. How should that be done?

a. $s_p^2 = \dfrac{9.3 + 9.8}{2}$

b. $s_p^2 = \dfrac{37(9.3) + 33(9.8)}{70}$

c. $s_p^2 = \dfrac{36(9.3) + 32(9.8)}{68}$

d. cannot be determined because the sample sizes are not the same for the two treatments

Use the following information for problems 42 and 43. A study has been conducted using a matched-pairs design. Thirty pairs were used in the study. The sample standard deviation under the first treatment is 12.7, and the sample standard deviation under the second treatment is 9.3. The standard deviation of the differences within each pair is 18.4.

42. What is the standard error of the estimated difference in treatment means ($\overline{X}_1 - \overline{X}_2$)?

a. $\sqrt{\dfrac{12.7}{30} + \dfrac{9.3}{30}}$

b. $\sqrt{\dfrac{12.7^2}{30} + \dfrac{9.3^2}{30}}$

c. $\sqrt{\dfrac{18.4}{30}}$

d. $\sqrt{\dfrac{18.4^2}{30}}$

43. How many degrees of freedom are associated with the standardized estimate of the difference in treatment means ($\overline{X}_1 - \overline{X}_2$)?

a. $30 - 1 = 29$

b. $30 + 30 - 2 = 58$

c. $30 + 30 = 60$

d. $\dfrac{\left(\dfrac{12.7^2}{30} + \dfrac{9.3^2}{30}\right)^2}{\dfrac{1}{29}\left(\dfrac{12.7^2}{30}\right)^2 + \dfrac{1}{29}\left(\dfrac{9.3^2}{30}\right)^2}$

Use the following information for problems 44, 45, and 46. A study has been conducted using a two-group design. Forty-eight units received the first treatment, and 57 received the second treatment. Based on theory, the researcher believes that the variances under the two treatments will be different. The sample standard deviation under the first treatment is 28.4 and under the second is 5.3.

44. What is the standard error of the estimated difference in treatment means ($\overline{X}_1 - \overline{X}_2$)?

a. $\sqrt{\dfrac{28.4}{48} + \dfrac{5.3}{57}}$

b. $\sqrt{\dfrac{28.4^2}{48} + \dfrac{5.3^2}{57}}$

b. $\dfrac{28.4^2}{\sqrt{48}} + \dfrac{5.3^2}{\sqrt{57}}$

c. $\dfrac{28.4}{48} + \dfrac{5.3}{57}$

45. How many degrees of freedom are associated with the standardized estimate of the difference in treatment means ($\overline{X}_1 - \overline{X}_2$)?

a. $48 + 57 - 2 = 103$

b. $48 + 57 = 105$

c. $\dfrac{\left(\dfrac{28.4}{48} + \dfrac{5.3}{57}\right)^2}{\dfrac{1}{47}\left(\dfrac{28.4}{48}\right)^2 + \dfrac{1}{56}\left(\dfrac{5.3}{57}\right)^2}$

d. $\dfrac{\left(\dfrac{28.4^2}{48} + \dfrac{5.3^2}{57}\right)^2}{\dfrac{1}{47}\left(\dfrac{28.4^2}{48}\right)^2 + \dfrac{1}{56}\left(\dfrac{5.3^2}{57}\right)^2}$

46. A chi-squared goodness-of-fit test was conducted. There were five categories, but no parameters were estimated. The value of the test statistic was 8.5. What is the p-value associated with the test?

a. $0.05 < p < 0.10$

b. $0.10 < p < 0.15$

c. $0.15 < p < 0.20$

d. $0.20 < p < 0.30$

Use the following information for problems 47 and 48. A researcher wanted to explore whether one gender was more likely than the other to be myopic (nearsighted). She took a random sample of 40 females and a random sample of 40 males from the students at a large university. She determined whether or not each selected individual was myopic. The results are in the following table:

MYOPIC?	FEMALE	MALE	TOTALS
Yes	26	19	45
No	14	21	35
Totals	40	40	80

47. What type of test is to be conducted?
   a. a paired t-test
   b. a chi-squared goodness-of-fit test
   c. a chi-squared test of homogeneity
   d. a chi-squared test of independence

48. How many degrees of freedom are associated with the test?
   a. 1
   b. 2
   c. 3
   d. 4

Use the following information to answer questions 49 and 50: A golf instructor thinks he has a new way of hitting the ball off of a tee that will make the ball go farther, but will not affect the variability in how far the ball is hit. He randomly selects 100 new golfers. He randomly assigns half of them to be taught the new method and half to be taught using the standard method. He does not tell any of them what he is doing. At the end of the month, he records how far each hits a ball off the first tee.

49. What is the appropriate statistical test for the hypotheses of interest?
   a. the z-test
   b. a paired t-test
   c. an independent t-test
   d. a chi-squared test of homogeneity

50. The value of the test statistic was 1.85. What is the p-value of the test and, using a 5% significance level, what is the conclusion?
   a. $0.25 < p < 0.05$. Do not reject the null hypothesis.
   b. $0.25 < p < 0.05$. Reject the null hypothesis.
   c. $0.05 < p < 0.1$. Do not reject the null hypothesis.
   d. $0.05 < p < 0.1$. Reject the null hypothesis.

▶ Answers

1. c
2. d
3. c
4. c
5. c
6. b
7. b
8. a
9. b
10. c
11. b
12. a
13. a
14. c
15. c
16. b
17. c
18. b
19. a
20. c
21. a
22. c
23. c
24. c
25. d
26. c
27. a
28. c
29. d
30. d
31. a
32. b
33. b
34. c
35. d
36. b
37. c
38. c
39. b
40. b
41. d
42. a
43. a
44. d
45. d
46. a
47. c
48. a
49. c
50. b

Answer Key

▶ Lesson 1

1. Population = all employees who work at the company; sample = the 200 employees randomly selected.
2. Population = all plants in the farmer's cotton field; sample = the 50 plants selected.
3. Population = the large shipment of bolts; sample = the 100 bolts selected.
4. Categorical
5. Numerical, continuous
6. Categorical
7. Numerical, discrete
8. Numerical, continuous
9. Categorical
10. Numerical, discrete

▶ Lesson 2

1. a. Only people with telephones could participate in the survey. People without telephones may be less likely to shop outside the city, causing the mean number of times people who shop outside the city to be underestimated.
   b. People might want to show support for their city, causing how often they shop outside of the city to be underestimated.
   c. People with lower incomes who can't afford to shop very often might be more reluctant to participate, causing the mean number of people who shop outside the city to be underestimated.

2. Measurement or response bias: She tends to undercount the birds farther from the track causing her to underestimate the average number of each kind within a ten-square yard area.

3. a. The cholesterol level of a participant after one month
   b. The people who participated in the study, narrow scope of inference
   c. Cause-and-effect conclusions can be made to determine the effect of the standard medication versus the effect of the recently developed medication
   d. Experiment, narrow scope of inference

4. a. Whether or not the customers would be willing to pay more
   b. All customers of the cable television company, broad scope of inference
   c. Associations can be concluded for the population as to whether or not people would be willing to pay more if a new set of channels was added.
   d. Sample survey

5. a. The number of correct responses from the students when shown the letters
   b. The students in the class, narrow scope of inference
   c. Cause-and-effect conclusions cannot be made.
   d. Observational study

6. a. The time required to clean the teeth
   b. All patients who have their teeth cleaned at this large dental school during this six-month period, broad scope of inference
   c. Cause-and-effect conclusions can be made.
   d. Experiment, broad scope of inference

▶ Lesson 3

1.

CATEGORY	FREQUENCY	RELATIVE FREQUENCY
Soda	31	0.37
Pop	27	0.33
Coke	25	0.30
Total	83	1.00

2.

CATEGORY	FREQUENCY	RELATIVE FREQUENCY
Had	267	.267
Had not	733	.733
Total	1,000	1.00

3.

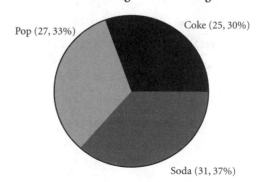

**Pie Chart of Word Used
When Ordering a Cola Beverage**

Pop (27, 33%)
Coke (25, 30%)
Soda (31, 37%)

4.

Pie Chart for Flu Vaccine

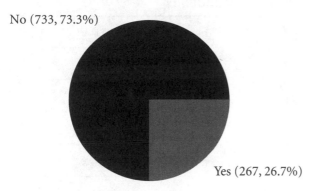

No (733, 73.3%)

Yes (267, 26.7%)

5.

Bar Chart of Word Used When Ordering Cola Beverage

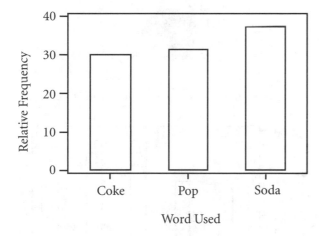

6.

Bar Chart of Flu Vaccine Distribution

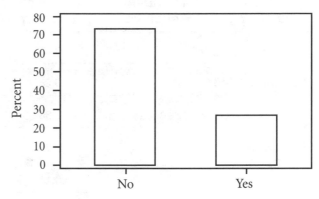

7. Explanatory variable: age and risk (healthy or at risk); Response variable: whether or not they received the flu vaccine.

8.

CATEGORY	FREQUENCY RECEIVED	RELATIVE FREQUENCY	FREQUENCY (DID NOT RECEIVE)	RELATIVE FREQUENCY
65+	146	0.63	87	0.37
High-Risk, 18–64	41	0.26	117	0.74
Healthcare Workers	23	0.36	41	0.64
Healthy, 18–48	24	0.07	318	0.93
Healthy, 50–64	33	0.16	170	0.84

9.

Flu Vaccine Pie Chart of People 65 and Older

Did Not Receive (.87, 37.3%)

Received (146, 62.7%)

Flu Vaccine Pie Chart for High-risk People Aged 18–64

Did Not Receive (117, 74.1%)

Received (41, 25.9%)

Flu Vaccine Pie Chart for Healthcare Workers

Did Not Receive (41, 64.1%)

Received (23, 35.9%)

Flu Vaccine Pie Chart for Healthy People Aged 18–48

Did Not Receive (318, 93.0%)

Received (24, 7.0%)

10.

Flu Vaccine Pie Chart for Healthy People Aged 50–84

Did Not Receive (170, 83.7%)

Received (33, 16.3%)

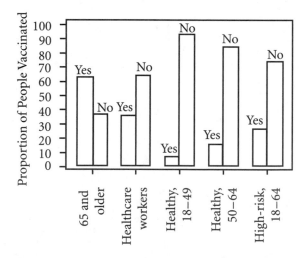

Bar Chart of Proportion of People Vaccinated

Age, Risk, and Health

► Lesson 4

1.

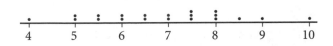

Average hours of sleep per night

The average hours of sleep ranged from four to ten hours. Most of the students averaged five to nine hours per night.

2.

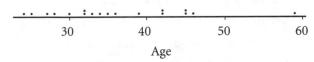

Age

The ages of the actors and actresses who won Oscars from 1996 to 2004 ranged from 24 to 54. The next oldest person to win was 46.

3.

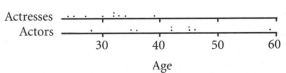

Age

4.

4	0
5	0 0 5 5
6	0 0 5 5
7	0 0 5 5 5
8	0 0 0 5
9	0
10	0

It appears that most students are getting between five and eight hours of sleep.

5.

2	4
2	5 7 8
3	0 2 2 3 4
3	5 6 9
4	2 2
4	5 5 6
5	
5	9

The most common ages at the time of receiving an Oscar award are from 25 to 40.

6.

FEMALES		MALES
4	2	8
7 5	2	
4 3 2 2 0	3	
9	3	5 6
	4	2 2
	4	5 5 6
	5	
	5	9

Female award winners tended to be younger than male award winners at the time they won their award. Both the male who was 28 and the one who was 59 seem to be unusual values.

▶ Lesson 5

1. All ages for actors and actresses who won an Oscar from 1997 to 2004 are included in the set of ages, not just some of the ages.
2. 30.7
3. 42
4. 32
5. Yes. There are two modes for the actors' ages: 42 and 45, which are two ages in the middle of the data set and close to the mean. The mode for the actress' ages is 32, which is exactly in the middle of the data set and close to the mean.
6. The median provides the best measure of central tendency because there are two outliers for the actors' ages.
7. The centers of distribution for the actors' ages were higher than for the actress' centers, showing that the actors were generally older than the actresses at the time they received their awards.
8. All students at the very large high school.
9. 6.9 hours
10. 7 hours
11. No. There are two modes: 7.5 and 8, both of which are slightly higher than the middle of the data set.
12. The mean would provide the best measure of center because the units of data are close in value with no outliers, but the median would also work well in this case.

▶ Lesson 6

1. 31
2. 15
3. 10
4. 7.5
5. 6.5
6. 3.7
7. Variance = 71.4; standard deviation = 8.4.
8. Variance = 20.0; standard deviation = 4.5.
9. The range, mean absolute deviation, and standard deviation are not good measures because they are affected by outliers. The only measure not influenced by outliers is the interquartile range, which would be the best measure to use.
10. Outliers affect the range, mean absolute deviation, and standard deviation, so the interquartile range would be the best measure because it is the only one not affected by outliers.
11. 6
12. 2.25
13. 1.2
14. Sample variance = 2.25; sample standard deviation = 1.5.
15. The range, mean absolute deviation, and standard deviation are affected by outliers, so the interquartile range would be the best measure of dispersion. It is the only one not affected by outliers.

▶ Lesson 7

1. Sample survey

2.

SHOE SIZE	FREQUENCY	RELATIVE FREQUENCY	CUMULATIVE RELATIVE FREQUENCY
7	2	0.04	0.04
8	6	0.12	0.16
9	14	0.28	0.44
10	13	0.26	0.70
11	11	0.22	0.92
12	4	0.08	1.00
Total	50	1.00	

3.

WAIT TIME (MINUTES)	FREQUENCY	RELATIVE FREQUENCY	CUMULATIVE RELATIVE FREQUENCY
0 – 1.9	21	0.42	0.42
2 – 3.9	11	0.22	0.64
4 – 5.9	10	0.20	0.84
6 – 7.9	1	0.02	0.86
8 – 9.9	3	0.06	0.92
10 – 11.9	3	0.06	0.98
12 – 13.9	0	0.00	0.98
14 – 15.9	0	0.00	0.98
16 – 17.9	0	0.00	0.98
18 – 19.9	1	0.02	1.00
Total	50	1.00	

4.

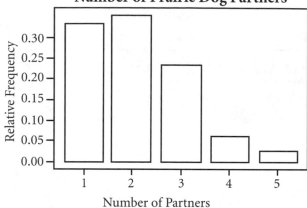

**Relative Frequency Histogram of
Number of Prairie Dog Partners**

6.

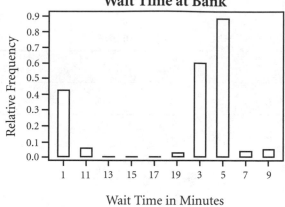

**Relative Frequency Histogram of
Wait Time at Bank**

5.

Histogram of Length of Stay in New Zealand

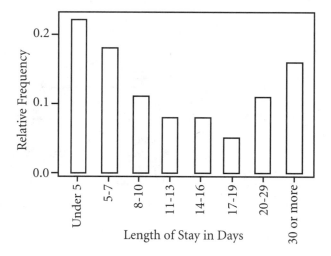

7.

Boxplot of Number of Partners

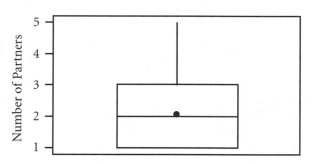

8.

Boxplot of Wait Time at the Bank

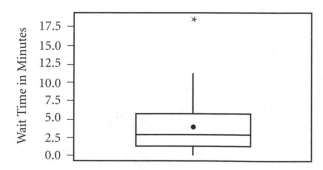

▶ Lesson 8

1. Explanatory variable = latitude; response variable = temperature.

2.

Scatter Plot of Average Temperature for the Month of March

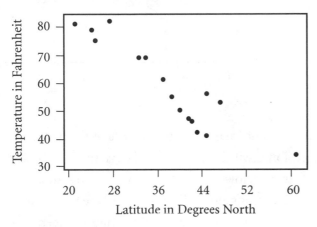

As the latitude increases, the average temperature decreases.

3.

$z_{Temperature}$	$z_{Latitude}$	$z_{Temperature}z_{Latitude}$
−1.61125	2.24775	−3.62169
−0.56700	0.20046	−0.11366
1.06465	−1.2618	−1.34346
0.15093	−0.09201	−0.01389
−0.17540	0.68791	−0.12066
0.67305	−0.51121	−0.34407
−1.15439	0.67816	−0.78286
−1.08912	0.61967	−0.67490
−0.82806	0.41494	−0.34360
−0.76279	0.37595	−0.28677
−0.24067	0.06398	−0.01540
1.52151	−1.01816	−1.54914
0.67305	−0.38448	−0.25877
−0.37120	0.92189	−0.34220
1.32571	−1.32038	−1.75044
1.39098	−1.62260	−2.25699
Total 0	0	−13.8185055

4. Pearson's correlation coefficient $r = -0.92$

5. $r = -0.92$ suggests a strong negative relationship between X and Y, latitude and temperature.

▶ Lesson 9

1. $S = \{1, 2, 3, 4\}$, where 1, 2, 3, and 4 each represent one of the four different movies showing at the theater.

2. There are 2^4 events: ϕ, {1}, {2}, {3}, {4}, {1,2}, {1,3}, {1,4}, {2,3}, {2,4}, {3,4}, {1,2,3}, {1,2,4}, {1,3,4}, {2,3,4}, S

3. {3}, {1,3}, {2,3}, {3,4}, {1,2,3}, {1,3,4}, {2,3,4}, S

4. No, because one movie might be more popular than another movie showing at the theater.

5. 0.70

6. 0.50

7. 0.30

8. 0.52

9. 0.44

10. 0.10

11. 0.43

12. No, $P(P) = \dfrac{495}{956} \neq \dfrac{95}{214} = P(P|Junior)$

13. 0.618

14. 0.379

▶ Lesson 10

1. 0.00390625

2. 0.0000000057

3. 0.0625

▶ Lesson 11

1. $\frac{1}{2}$

2. 0.7734

3. 0.2266

4. 0.0287

5. 0.5647

6. 0.08

7. 0.61

8. −1.88

9. 0.2266

10. 6.17

▶ Lesson 12

1. Statistic

2. Parameter

3. μ

4. $\overline{X}$

5. 73 years

6. 1.5 years

7. For 95% of all randomly selected samples of size $n = 75$ residents, the mean age will be between 70 and 76 years.

8. $t^* = 2.473$

9. $t^* = 2.807$

10. $t^* = -2.528$

▶ Lesson 13

1.

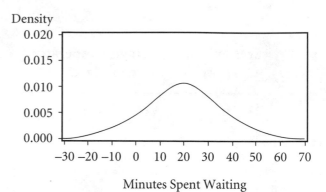

2. For a normal distribution, 95% of the population is within two standard deviations of the mean. However, in this example, two standard deviations either direction from the mean, we have that 95% of the population is between −10 and 50. It is impossible to wait a negative number of hours. Thus, the distribution is most likely skewed right.

3. By the Law of Large Numbers, the sample mean of 20 is likely to be "close" to the population mean. (A sample size of 100 is usually considered to be large.)

4. Normal with mean μ and standard deviation $\frac{\sigma}{\sqrt{n}}$

▶ Lesson 14

1. Point estimate = 0.83; interval of reasonable values is between 0.79 and 0.87.

2. Stratified random sample

3. Cluster sample

ANSWER KEY

▶ Lesson 15

1. Confidence interval = (0.57, 0.73)
2. Response bias; people who didn't pass could have been embarrassed and said they did pass.
3. A 99% confidence interval = (0.56, 0.76), compared to a 98% confidence interval of (0.57, 0.73). Clearly, the 99% confidence interval is wider. This makes sense because, as the interval increases in length, we become more confident that the interval will capture the true population proportion.
4. A 99% confidence interval = (0.54, 0.76). As the sample size decreased, the confidence interval widened.

▶ Lesson 16

1. H_0 = do not go to the doctor; H_a = go to the doctor.
2. A type I error would occur if the woman did go to the doctor when it really wasn't necessary, thus needlessly spending money.
3. A type II error would occur if the woman did not go to the doctor when she was really in need of medical attention, possibly causing her illness to get worse.
4. (i) The set of hypotheses to be tested are H_0: $p = 0.70$ versus H_a: $p > 0.70$. (ii) The mayor took a random sample of adult residents of the city, so the first condition for inference is satisfied. To check the second condition, we have $np = 150(\frac{108}{150}) = 108 > 10$ and $n(1 - p) = 150(1 - \frac{108}{150}) = 42 > 10$. Thus, the second condition for inference is also satisfied. The test statistic = 0.53. (iii) The p-value is 0.2981. (iv) We do not reject the null hypothesis. (v) There is not suffi-

cient evidence to conclude that more than 70% of the city's adult residents favor a ban on smoking in public places.

▶ Lesson 17

1. (3.44, 3.50)
2. We are 90% confident that the mean nitrogen level of lakes in this state is within 4.2 μg/L of 754 μg/L within the study region.

▶ Lesson 18

1. All patients would have one cream applied to one of their hands and the other cream applied to the other hand. Which hand the creams are applied to would be randomly decided.
2. The dermatologist would randomly select 20 of her patients to use one type of cream on both of their hands; the other 20 patients would use the other type of cream on both of their hands.
3. The matched-pairs design would be best for this study. People vary in the types of jobs and activities they perform in a day, causing some people to use their hands more than others. This might affect the results. With the matched-pairs design we can eliminate this source of variation.
4. The two treatments are liquid A and liquid B. Each treatment is given to a student. Thus, liquid A and liquid B are paired by the student. A pair consists of the boiling points for the two liquids given to each student. Which side of the hot plate would be used to boil liquids A and B was determined randomly.

5.

Student	1	2	3	4	5
Difference	8.5	6.8	6.5	6.6	7.5
Student	6	7	8	9	10
Difference	8.8	6.9	5.8	8.7	8.0
Student	11	12	13	14	15
Difference	7.1	6.4	6.8	8.4	5.5

6. Estimated mean = 7.22; estimated standard deviation = 1.050

7. 0.271

8.

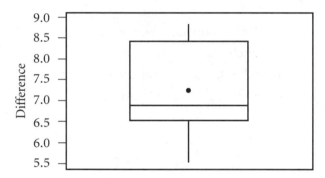

Boxplot of Difference in Boiling Temperatures

Dotplot of Difference in Boiling Temperatures

Because there are very few data values, it is difficult to fully asses normality. However, looking at the data we have, we see that there may be some hint of skew to the right, but over all it is fairly symmetric. Therefore, it is not unreasonable to assume the data are normally distributed.

9. It depends on the source of saltwater and plain water. For example, if the teacher had a gallon of distilled water and used it to provide all the plain water and to mix the saltwater, inference can only be made to the water used in the study.

10. We estimate that, on average, liquid A boiled at 7.22 degrees Celsius higher than liquid B, and we are 95% confident that this estimate is within 0.58 degrees of the true mean difference in the temperatures at which both liquids began to boil.

11. The test statistic = 26.64 and the p-value < $2(0.001) = 0.002$. The mean temperature at which liquid B began to boil was significantly less than the mean temperature at which liquid A began to boil. From this we can guess that liquid A is saltwater and liquid B is plain water.

▶ **Lesson 19**

1. Half of the students were randomly chosen to be given caffeine and the other half were randomly assigned not to receive caffeine.

2. The mean and standard deviation for students given caffeine are estimated to be 11.73 and 1.87, respectively. The mean score for students who did not receive caffeine is estimated to be 11.65 with a standard deviation of 2.52.

3. The standard deviations for the two treatment groups are very close, therefore it is reasonable to assume that they are estimating a common variance. The estimate of that common variance = 4.92 points^2.

4. The estimated difference in the mean score of a memory test taken by students who had consumed caffeine and students who had not is 0.08. The standard error of this estimate is 0.44 points.

5.

Dotplots of Memory Test Scores

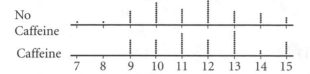

Boxplots of Memory Test Scores

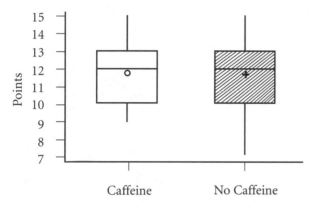

Because the data are discrete, it is impossible for the memory scores to be normally distributed. However, the sample size is large enough that the sampling distribution of $\overline{X}$ can reasonably be assumed to be normal by the Central Limit Theorem.

6. The group of students from which the teacher made his random selection

7. We estimate that the mean of students' scores who consumed caffeine is 0.08 points greater than the mean of students' scores who did not have caffeine, and we are 95% confident this estimate is within 0.88 points of the difference in these two population means. There is such a small difference in the population means that not much support seems to exist for student's belief that ingesting caffeine while studying improves memory.

▶ **Lesson 20**

1. The response variable of interest is the customer's preferred sandwich choice. We may define

p_T = proportion of customer's ordering turkey
p_B = proportion of customer's ordering beef
p_V = proportion of customer's ordering veggie
p_S = proportion of customer's ordering salami
p_H = proportion of customer's ordering ham
p_P = proportion of customer's ordering pastrami

2. The set of hypotheses of interest is

H_0: $p_T = 0.28$, $p_B = 0.12$, $p_V = 0.20$, $p_S = 0.14$, $p_H = 0.22$, $p_P = 0.04$

H_a: not H_0

3. p-value $= 0.0250$

4. The p-value is between 0.05 and 0.1.

5. p-value < 0.0005

6. p-value > 0.3

7.

SANDWICH TYPE	OBSERVED	EXPECTED	$\frac{(n_i - e_i)^2}{e_i}$
Turkey	45	42	0.2143
Beef	14	18	0.8889
Veggie	33	30	0.3000
Salami	16	21	1.1905
Ham	37	33	0.4848
Pastrami	5	6	0.1667
Totals	150	150	3.2452

8. 3.2452

9. p-value > 0.30

10. We would not reject the null hypothesis that the percentages of turkey, beef, veggie, salami, ham, and pastrami being ordered are as the sandwich shop owner has hypothesized.

11. $H_0 : p_{1j} = p_1, p_{2j} = p_2, \ldots p_{rj}, j = M, F$ where M stands for males and F stands for females.
H_a : not H_0.

12.

TYPE OF VEHICLE	MALE	FEMALE	TOTAL
Trucks	30	30	60
Vans	17.5	17.5	35
SUVs	27.5	27.5	55
Convertibles	8.5	8.5	17
Sedans	31.5	31.5	63
Other	10	10	20
Totals	125	125	250

13. The conditions for the test are satisfied. The test statistic = 25.2554.

14. p-value < 0.0005

15. We reject the null hypothesis.

16. Evidence exists that males and females differ in the proportion of the types of vehicles they drive

17. $X^2 = 36.9370$; $p < 0.001$. We conclude to reject the null hypothesis. A relationship exists between whether or not a student lives on campus and how much school spirit he or she has.

Appendix: How to Prepare for a Test ▶

A standardized test is nothing to fear. Many people clutch and worry about a testing situation, but you're much better off taking that nervous energy and turning it into something positive that will help you do well on your test rather than inhibit your testing ability. The following pages include valuable tips for combating test anxiety, that sinking or blank feeling some people get as they begin a test or encounter a difficult question. Next, you will find valuable tips for using your time wisely and for avoiding errors in a testing situation. Finally, you will find a plan for preparing for the test, a plan for the test day, and a suggestion for an after-test activity.

▶ Combating Test Anxiety

Knowing what to expect and being prepared for it is the best defense against test anxiety, that worrisome feeling that keeps you from doing your best. Practice and preparation keeps you from succumbing to that feeling. Nevertheless, even the brightest, most well-prepared test takers may suffer from occasional bouts of test anxiety. But don't worry; you can overcome it.

Take the Test One Question at a Time

Focus all of your attention on the one question you're answering. Block out any thoughts about questions you've already read or concerns about what's coming next. Concentrate your thinking where it will do the most good—on the question you're answering.

Develop a Positive Attitude

Keep reminding yourself that you're prepared. You've studied hard, so you're probably better prepared than most others who are taking the test. Remember, it's only a test, and you're going to do your best. That's all anyone can ask of you. If that nagging drill sergeant inside your head starts sending negative messages, combat him or her with positive ones of your own.

- "I'm doing just fine."
- "I've prepared for this test."
- "I know exactly what to do."
- "I know I can get the score I'm shooting for."

You get the idea. Remember to drown out negative messages with positive ones of your own.

If You Lose Your Concentration

Don't worry about it! It's normal. During a long test, it happens to everyone. When your mind is stressed or overexerted, it takes a break whether you want it to or not. It's easy to get your concentration back if you simply acknowledge the fact that you've lost it and take a quick break. Your brain needs very little time (seconds really) to rest.

Put your pencil down and close your eyes. Take a few deep breaths and listen to the sound of your breathing. The ten seconds or so that this takes is really all the time your brain needs to relax and get ready to focus again.

Try this technique several times in the days before the test when you feel stressed. The more you practice, the better it will work for you on the day of the test.

If You Freeze before or during the Test

Don't worry about a question that stumps you even though you're sure you know the answer. Mark it and go on to the next question. You can come back to the stumper later. Try to put it out of your mind completely until you come back to it. Just let your subconscious chew on the question while your conscious mind focuses on the other items (one at a time, of course). Chances are, the memory block will be gone by the time you return to the question.

If you freeze before you begin the test, here's what to do:

1. Take a little time to look over the test.
2. Read a few of the questions.
3. Decide which ones are the easiest and start there.
4. Before long, you'll be "in the groove."

▶ Time Strategies

Use your time wisely to avoid making careless errors.

Pace Yourself

The most important time strategy is to pace yourself. Before you begin, take just a few seconds to survey the test, making note of the number of questions and of the sections that look easier than the rest. Rough out a time schedule based upon the amount of time available to you. Mark the halfway point on your test and make a note beside that mark of what the time will be when the testing period is half over.

Keep Moving

Once you begin the test, keep moving. If you work slowly in an attempt to make fewer mistakes, your mind will become bored and begin to wander. You'll end up making far more mistakes if you're not concentrating.

As long as we're talking about mistakes, don't stop for difficult questions. Skip them and move on. You can come back to them later if you have time. A question that takes you five seconds to answer counts as much as one that takes you several minutes, so pick up the easy points first. Besides, answering the easier questions first helps to build your confidence and gets you in the testing groove. Who knows? As you go through the test, you may even stumble across some relevant information to help you answer those tough questions.

Don't Rush

Keep moving, but don't rush. Think of your mind as a seesaw. On one side is your emotional energy. On the other side is your intellectual energy. When your emotional energy is high, your intellectual capacity is low. Remember how difficult it is to reason with someone when you're angry? On the other hand, when your intellectual energy is high, your emotional energy is low. Rushing raises your emotional energy. Remember the last time you were late for work? All that rushing around caused you to forget important things—like your lunch. Move quickly to keep your mind from wandering, but don't rush and get flustered.

Check Yourself

Check yourself at the halfway mark. If you're a little ahead, you know you're on track and may even have a little time left to check your work. If you're a little behind, you have several choices. You can pick up the pace a little, but do this only if you can do it comfortably. Remember—don't rush! You can also skip around in the remaining portion of the test to pick up as many easy points as possible. This strategy has one drawback, however. If you are marking a bubble-style

answer sheet, and you put the right answers in the wrong bubbles—they're wrong. So pay close attention to the question numbers if you decide to do this.

▶ Avoiding Errors

When you take the test, you want to make as few errors as possible in the questions you answer. Here are a few tactics to keep in mind.

Control Yourself

Remember the comparison between your mind and a seesaw that you read a few paragraphs ago? Keeping your emotional energy low and your intellectual energy high is the best way to avoid mistakes. If you feel stressed or worried, stop for a few seconds. Acknowledge the feeling (Hmmm! I'm feeling a little pressure here!), take a few deep breaths, and send yourself a few positive messages. This relieves your emotional anxiety and boosts your intellectual capacity.

Directions

In many standardized testing situations, a proctor reads the instructions aloud. Make certain you understand what is expected. If you don't, ask. Listen carefully for instructions about how to answer the questions and make certain you know how much time you have to complete the task. Write the time on your test if you don't already know how long you have to take the test. If you miss this vital information, ask for it. You need it to do well on your test.

Answers

Place your answers in the right blanks or the corresponding ovals on the answer sheet. Right answers in the wrong place earn no points. It's a good idea to check every five to ten questions to make sure you're in the right spot. That way you won't need much time to correct your answer sheet if you have made an error.

▶ Reading Long Passages

Frequently, standardized tests are designed to test your reading comprehension. The reading sections often contain passages of a paragraph or more. Here are a few tactics for approaching these sections.

This may seem strange, but some questions can be answered without ever reading the passage. If the passage is short, a paragraph around four sentences or so, read the questions first. You may be able to answer them by using your common sense. You can check your answers later after you've actually read the passage. Even if you can't answer any of the questions, you know what to look for in the passage. This focuses your reading and makes it easier for you to retain important information. Most questions will deal with isolated details in the passage. If you know what to look for ahead of time, it's easier to find the information.

If a reading passage is long and is followed by more than ten questions, you may end up spending too much time reading the questions first. Even so, take a few seconds to skim the questions and read a few of the shorter ones. As you read, mark up the passage. If you find a sentence that seems to state the main idea of the passage, underline it. As you read through the rest of the passage, number the main points that support the main idea. Several questions will deal with this information. If it's underlined and numbered, you can locate it easily. Other questions will ask for specific details. Circle information that tells who, what, when, or where. The circles will be easy to locate later if you run across a question that asks for specific information. Marking up a passage in this way also heightens your concentration and makes it more likely that you'll

remember the information when you answer the questions following the passage.

Choosing the Right Answers

Make sure you understand what the question is asking. If you're not sure of what's being asked, you'll never know whether you've chosen the right answer. So figure out what the question is asking. If the answer isn't readily apparent, look for clues in the answer choices. Notice the similarities and differences in the answer choices. Sometimes, this helps to put the question in a new perspective and makes it easier to answer. If you're still not sure of the answer, use the process of elimination. First, eliminate any answer choices that are obviously wrong. Then reason your way through the remaining choices. You may be able to use relevant information from other parts of the test. If you can't eliminate any of the answer choices, you might be better off to skip the question and come back to it later. If you can't eliminate any answer choices to improve your odds when you come back later, then make a guess and move on.

If You're Penalized for Wrong Answers

You must know whether there's a penalty for wrong answers before you begin the test. If you don't, ask the proctor before the test begins. Whether you make a guess or not depends upon the penalty. Some standardized tests are scored in such a way that every wrong answer reduces your score by one-fourth or one-half of a point. Whatever the penalty, if you can eliminate enough choices to make the odds of answering the question better than the penalty for getting it wrong, make a guess.

Let's imagine you are taking a test in which each answer has four choices and you are penalized one-fourth of a point for each wrong answer. If you have no clue and cannot eliminate any of the answer choices, you're better off leaving the question blank because the odds of answering correctly are one in four. This makes the penalty and the odds equal. However, if you can eliminate one of the choices, the odds are now in your favor. You have a one in three chance of answering the question correctly. Fortunately, few tests are scored using such elaborate means, but if your test is one of them, know the penalties and calculate your odds before you take a guess on a question.

If You Finish Early

Use any time you have left at the end of the test or test section to check your work. First, make certain you've put the answers in the right places. As you're doing this, make sure you've answered each question only once. Most standardized tests are scored in such a way that questions with more than one answer are marked wrong. If you've erased an answer, make sure you've done a good job. Check for stray marks on your answer sheet that could distort your score.

After you've checked for these obvious errors, take a second look at the more difficult questions. You've probably heard the folk wisdom about never changing an answer. If you have a good reason for thinking a response is wrong, change it.

▶ The Days before the Test

To do your very best on an exam, you have to take control of your physical and mental state. Exercise, proper diet, and rest will ensure that your body works with, rather than against, your mind on exam day, as well as during your preparation.

Physical Activity

Get some exercise in the days preceding the test. You'll send some extra oxygen to your brain and allow your thinking performance to peak on the day you take the test. Moderation is the key here. You don't want to exercise so much that you feel exhausted, but a little physical activity will invigorate your body and brain.

Balanced Diet

Like your body, your brain needs the proper nutrients to function well. Eat plenty of fruits and vegetables in the days before the test. Foods that are high in lecithin, such as fish and beans, are especially good choices. Lecithin is a mineral your brain needs for peak performance. You may even consider a visit to your local pharmacy to buy a bottle of lecithin tablets several weeks before your test.

Rest

Get plenty of sleep the nights before you take the test. Don't overdo it though or you'll make yourself as groggy as if you were overtired. Go to bed at a reasonable time, early enough to get the number of hours you need to function effectively. You'll feel relaxed and rested if you've gotten plenty of sleep in the days before you take the test.

Trial Run

At some point before you take the test, make a trial run to the testing center to see how long it takes. Rushing raises your emotional energy and lowers your intellectual capacity, so you want to allow plenty of time on the test day to get to the testing center. Arriving 10 or 15 minutes early gives you time to relax and get situated.

Test Day

It's finally here, the day of the big test. Set your alarm early enough to allow plenty of time. Eat a good breakfast. Avoid anything that's really high in sugar, such as donuts. A sugar high turns into a sugar low after an hour or so. Cereal and toast, or anything with complex carbohydrates is a good choice. Eat only moderate amounts. You don't want to take a test feeling stuffed!

Pack a high-energy snack to take with you. You may have a break sometime during the test when you can grab a quick snack. Bananas are great. They have a moderate amount of sugar and plenty of brain nutrients, such as potassium. Most proctors won't allow you to eat a snack while you're testing, but a peppermint shouldn't pose a problem. Peppermints are like smelling salts for your brain. If you lose your concentration or suffer from a momentary mental block, a peppermint can get you back on track. Don't forget the earlier advice about relaxing and taking a few deep breaths.

Leave early enough so you have plenty of time to get to the test center. Allow a few minutes for unexpected traffic. When you arrive, locate the restroom and use it. Few things interfere with concentration as much as a full bladder. Then find your seat and make sure it's comfortable. If it isn't, tell the proctor and ask to change to something you find more suitable.

Now relax and think positively! Before you know it the test will be over, and you'll walk away knowing you've done as well as you can.

After the Test

Two things are important for after the test:

1. Plan a little celebration.
2. Go to it.

If you have something to look forward to after the test is over, you may find it easier to prepare well for the test and to keep moving during the test. Good luck!